OFFICIAL SQA PAST PAPERS WITH ANSWERS

HIGHER

HUMAN BIOLOGY
2008-2012

2008 EXAM — page 3
2009 EXAM — page 33
2010 EXAM — page 67
2011 EXAM — page 101
2012 EXAM — page 135
ANSWER SECTION — page 173

SQA

BrightRED
PUBLISHING

First exam published in 2008.
Published by Bright Red Publishing Ltd, 6 Stafford Street, Edinburgh EH3 7AU
tel: 0131 220 5804 fax: 0131 220 6710 info@brightredpublishing.co.uk www.brightredpublishing.co.uk

ISBN 978-1-84948-292-9

A CIP Catalogue record for this book is available from the British Library.

Bright Red Publishing is grateful to the copyright holders, as credited on the final page of the Question Section, for permission to use their material. Every effort has been made to trace the copyright holders and to obtain their permission for the use of copyright material. Bright Red Publishing will be happy to receive information allowing us to rectify any error or omission in future editions.

HIGHER

2008

[BLANK PAGE]

FOR OFFICIAL USE

Total for
Sections B & C

X009/301

NATIONAL
QUALIFICATIONS
2008

TUESDAY, 27 MAY
1.00 PM – 3.30 PM

HUMAN BIOLOGY
HIGHER

Fill in these boxes and read what is printed below.

Full name of centre

Town

Forename(s)

Surname

Date of birth

Day Month Year   Scottish candidate number   Number of seat

SECTION A–Questions 1—30

Instructions for completion of Section A are given on page two.

For this section of the examination you must use an **HB pencil**.

SECTIONS B AND C

1   (a) All questions should be attempted.

      (b) It should be noted that in **Section C** questions 1 and 2 each contain a choice.

2   The questions may be answered in any order but all answers are to be written in the spaces provided in this answer book, **and must be written clearly and legibly in ink**.

3   Additional space for answers will be found at the end of the book. If further space is required, supplementary sheets may be obtained from the invigilator and should be inserted inside the **front** cover of this book.

4   The numbers of questions must be clearly inserted with any answers written in the additional space.

5   Rough work, if any should be necessary, should be written in this book and then scored through when the fair copy has been written. If further space is required a supplementary sheet for rough work may be obtained from the invigilator.

6   Before leaving the examination room you must give this book to the invigilator. If you do not, you may lose all the marks for this paper.

Read carefully

1 Check that the answer sheet provided is for **Human Biology Higher (Section A)**.

2 For this section of the examination you must use an **HB pencil**, and where necessary, an eraser.

3 Check that the answer sheet you have been given has **your name**, **date of birth**, **SCN** (Scottish Candidate Number) and **Centre Name** printed on it.

   Do not change any of these details.

4 If any of this information is wrong, tell the Invigilator immediately.

5 If this information is correct, **print** your name and seat number in the boxes provided.

6 The answer to each question is **either** A, B, C or D. Decide what your answer is, then, using your pencil, put a horizontal line in the space provided (see sample question below).

7 There is **only one correct** answer to each question.

8 Any rough working should be done on the question paper or the rough working sheet, **not** on your answer sheet.

9 At the end of the exam, put the **answer sheet for Section A inside the front cover of this answer book**.

Sample Question

The digestive enzyme pepsin is most active in the

A   stomach

B   mouth

C   duodenum

D   pancreas.

The correct answer is **A**—stomach. The answer **A** has been clearly marked in **pencil** with a horizontal line (see below).

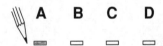

Changing an answer

If you decide to change your answer, carefully erase your first answer and, using your pencil, fill in the answer you want. The answer below has been changed to **D**.

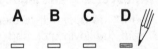

SECTION A

All questions in this section should be attempted.

Answers should be given on the separate answer sheet provided.

1. During which of the following chemical conversions is ATP produced?

   A   Amino acids ⟶ protein

   B   Glucose ⟶ pyruvic acid

   C   Haemoglobin ⟶ oxyhaemoglobin

   D   Nucleotides ⟶ mRNA

2. The following statements relate to respiration and the mitochondrion.

   1   Glycolysis takes place in the mitochondrion.

   2   The mitochondrion has two membranes.

   3   The rate of respiration is affected by temperature.

   Which of the above statements are correct?

   A   1 and 2

   B   1 and 3

   C   2 and 3

   D   All of them

3. The anaerobic breakdown of glucose splits from the aerobic pathway of respiration

   A   after the formation of pyruvic acid

   B   after the formation of acetyl CoA

   C   after the formation of citric acid

   D   at the start of glycolysis.

4. In respiration, the products of the cytochrome system are

   A   hydrogen and carbon dioxide

   B   water and ATP

   C   oxygen and ADP

   D   pyruvic acid and water.

5. The key below can be used to identify carbohydrates.

   1   Soluble....................................... 2
       Insoluble.................................. glycogen

   2   Benedict's test positive 3
       Benedict's test negative............ sucrose

   3   Barfoed's test positive 4
       Barfoed's test negative .... lactose

   4   Clinistix test positive .... glucose
       Clinistix test negative .... fructose

   Which line in the table of results below is **not** in agreement with the information contained in the key?

	Carbohydrate	Benedict's test	Barfoed's test	Clinistix test
A	lactose	positive	negative	not tested
B	glucose	positive	negative	positive
C	fructose	positive	positive	negative
D	sucrose	negative	not tested	not tested

6. Which of the following is an immune response?

   A   T-lymphocytes secreting antigens

   B   T-lymphocytes carrying out phagocytosis

   C   B-lymphocytes combining with foreign antigens

   D   B-lymphocytes producing antibodies

7. Phagocytes contain many lysosomes so that

   A   enzymes which destroy bacteria can be stored

   B   toxins from bacteria are neutralised

   C   antibodies can be released in response to antigens

   D   bacteria can be engulfed into the cytoplasm.

8. Which of the following is an example of active immunity?

   A   Antibody production following exposure to antigens

   B   Antibodies crossing the placenta from mother to fetus

   C   Antibodies passing from the mother's milk to a suckling baby

   D   Antibody extraction from one mammal to inject into another

9. The following steps occur during the replication of a virus.

   1   Alteration of host's cell metabolism

   2   Production of viral protein coats

   3   Replication of viral DNA

   In which sequence do these events occur?

   A   1 → 3 → 2

   B   1 → 2 → 3

   C   2 → 1 → 3

   D   3 → 1 → 2

10. The diagram below shows a stage in meiosis.

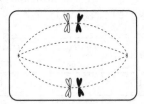

   Which of the following diagrams shows the next stage in meiosis?

   A   B

   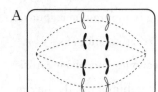   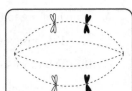

   C   D

   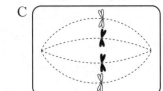   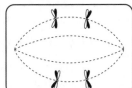

11. Cystic fibrosis is a genetic condition caused by an allele which is not sex-linked.

   A child is born with cystic fibrosis despite neither parent having the condition.

   The parents are going to have a second child. What is the chance this child will have cystic fibrosis?

   A   75%

   B   67%

   C   50%

   D   25%

12. A sex-linked condition in humans is caused by a recessive allele. What is the chance of an unaffected man and a carrier woman having an unaffected male child?

   A   1 in 1

   B   1 in 2

   C   1 in 3

   D   1 in 4

13. One function of the seminal vesicles is to

   A   produce testosterone

   B   allow sperm to mature

   C   store sperm temporarily

   D   produce nutrients for sperm.

14. Which fertility treatment would be appropriate for a woman with blocked uterine tubes?

   A   Provision of fertility drugs

   B   *In vitro* fertilisation

   C   Artificial insemination

   D   Calculation of fertile period

15. A 30 g serving of breakfast cereal contains 1·5 mg of iron. Only 25% of this iron is absorbed into the bloodstream.

   If a pregnant woman requires 6 mg of iron per day, how much cereal would she have to eat each day to meet this requirement?

   A   60 g

   B   120 g

   C   240 g

   D   480 g

16. Which of the following blood vessels carries oxygenated blood?

   A   Renal vein

   B   Hepatic vein

   C   Pulmonary vein

   D   Hepatic portal vein

17. In which of the following situations might a fetus be at risk from Rhesus antibodies produced by the mother?

	Father	Mother
A	Rhesus positive	Rhesus negative
B	Rhesus positive	Rhesus positive
C	Rhesus negative	Rhesus negative
D	Rhesus negative	Rhesus positive

18. The diagram below shows an ECG trace taken during exercise.

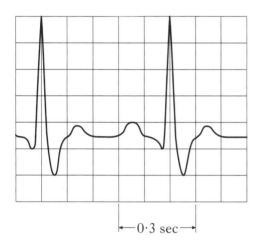

←—0·3 sec—→

The person's heart rate is

A   80 bpm

B   100 bpm

C   120 bpm

D   140 bpm.

19. The diagram below shows a section through the human heart.

What is the correct position of the pacemaker?

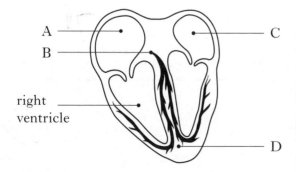

20. The vessel by which blood leaves the liver is the

A   renal vein

B   hepatic portal vein

C   renal artery

D   hepatic vein.

21. The graph below shows an individual's skin temperature and rate of sweat production over a period of 50 minutes.

Key

——— sweat production

------ skin temperature

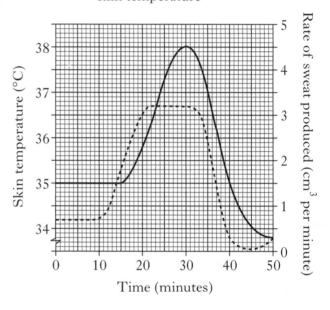

What is the skin temperature when the rate of sweat production is at a maximum?

A   3·2 °C

B   4·5 °C

C   36·7 °C

D   38·0 °C

[Turn over

22. The following diagram represents four neurones in a neural pathway.

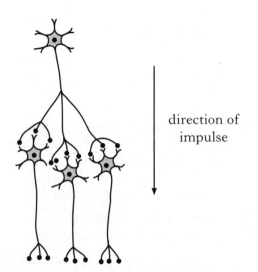

direction of impulse

Which line in the table describes the pathway correctly?

	Type of pathway	
A	motor	divergent
B	motor	convergent
C	sensory	divergent
D	sensory	convergent

23. Which of the following carries an impulse towards a nerve cell body?

A   Dendrite

B   Axon

C   Myelin

D   Myosin

24. Which of the following statements describes a neurotransmitter and its method of removal?

A   Adrenaline is removed by reabsorption.

B   Adrenaline is removed by enzyme degradation.

C   Noradrenaline is removed by enzyme degradation.

D   Noradrenaline is removed by reabsorption.

25. The diagram below illustrates the relationship between short and long-term memory.

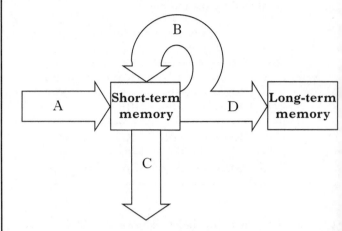

Which arrow represents the process of rehearsal?

26. The behavioural term *generalisation* is defined correctly as the ability to

A   make appropriate different responses to different but related stimuli

B   make the same appropriate response to different but related stimuli

C   submerge one's personal identity in the anonymity of a group

D   improve performance in competitive situations.

27. The table below contains information about the populations of four countries in the year 2000.

In which country did the population decrease?

	Number per 1000 inhabitants			
Country	Births	Deaths	Immigrants	Emigrants
A	9·3	10·1	1·0	0·1
B	9·7	10·3	1·3	0·4
C	10·1	9·9	0·2	0·5
D	10·8	10·5	0·1	0·3

28. The diagram below shows the number of people dying from different causes in a developing country. (Figures are in millions.)

Causes of death

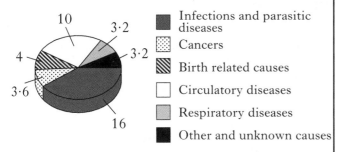

What percentage of deaths is due to birth related causes?

A   4%

B   8%

C   10%

D   11%

29. Which of the following processes is carried out by bacteria found in root nodules?

A   Denitrification

B   Nitrification

C   Nitrogen fixation

D   Deamination

30. Which of the following does **not** play a part in global warming?

A   The cutting down of forests

B   Methane production by cattle

C   The increase in use of motor vehicles

D   The increased use of fertilisers on farmland

Candidates are reminded that the answer sheet MUST be returned INSIDE the front cover of this answer booklet.

[Turn over for Section B

DO NO
WRITE
THIS
MARG

Marks

SECTION B

All questions in this section should be attempted.

All answers must be written clearly and legibly in ink.

1.   The diagram below illustrates the two main stages of protein synthesis.

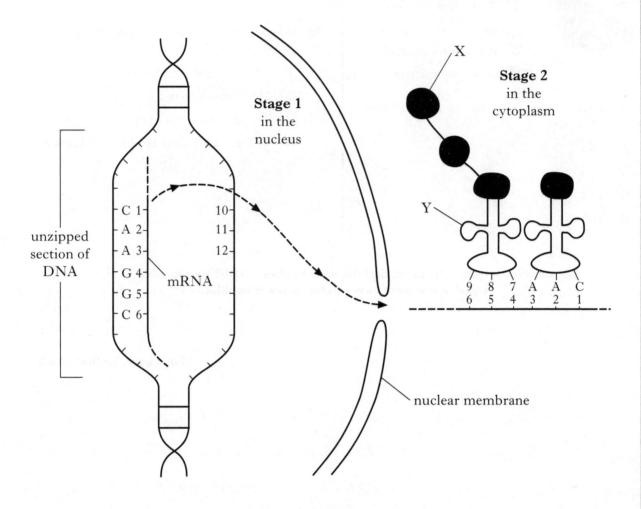

(a)   Describe **three** differences between DNA and mRNA.

1 _____

_____

2 _____

_____

3 _____

_____   2

DO NOT WRITE IN THIS MARGIN

1. (continued) *Marks*

(*b*) Name bases 3, 8 and 11.

Base 3 ______________________

Base 8 ______________________

Base 11 ______________________ **2**

(*c*) **Circle** a codon in the diagram opposite. **1**

(*d*) Where in the cytoplasm does stage 2 take place?

______________________ **1**

(*e*) Name molecules X and Y.

X ______________________ Y ______________________ **1**

(*f*) The newly synthesised protein may be secreted from the cell.

(i) Name the cell structure where the protein would be found just before it enters a secretory vesicle.

______________________ **1**

(ii) Describe what happens to the protein while it is in this cell structure.

______________________ **1**

[Turn over

2. (*a*) The diagram below shows some of the functions of proteins in the cell membrane.

Marks

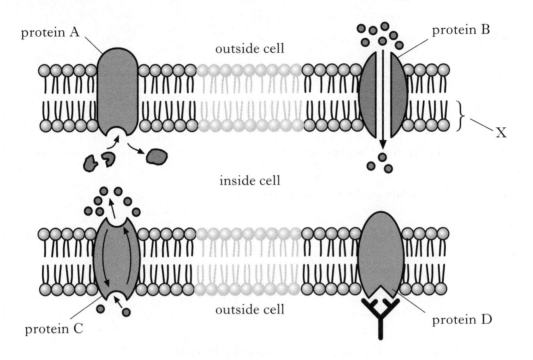

(i) Use the information from the diagram to complete the table below.

Protein	Function
	Transports molecules by diffusion
A	
D	
	Transports molecules by active transport

3

(ii) Identify molecule X and describe its function within the membrane.

Molecule X _____

Function _____

_____

2

(*b*) Describe what happens to the cell membrane during the process of endocytosis.

_____

_____

_____

2

DO NOT
WRITE IN
THIS
MARGIN

Marks

3. The blood group of an individual is controlled by three alleles *A*, *B* and *O*.

Alleles *A* and *B* are co-dominant and completely dominant to allele *O*.

The diagram below shows the blood groups of three generations of a family.

Parents   Mother   Father
               Group B   Group A

Children   Son 1   Son 2   Daughter ———— Husband
               Group A   Group O   Group O

Grandchildren   Grandson   Granddaughter
                                     Group B   Group A

(a) What is the blood group of the daughter?

_____   1

(b) State the genotypes of the grandchildren.

Grandson _____   Granddaughter _____   1

(c) How many of the three children are homozygous?

_____   1

(d) Explain the meaning of the term *co-dominant*.

_____

_____   1

(e) Only one of the sons can safely receive a blood transfusion from his brother.
Indicate whether this statement is true or false and explain your decision.

True/False _____

Explanation _____

_____

_____

_____   2

[Turn over

4. The graphs below show the plasma concentrations of certain hormones throughout a woman's menstrual cycle.

   Graph 1 shows the concentrations of FSH and LH.

   Graph 2 shows the concentration of two other hormones, X and Y.

Graph 1

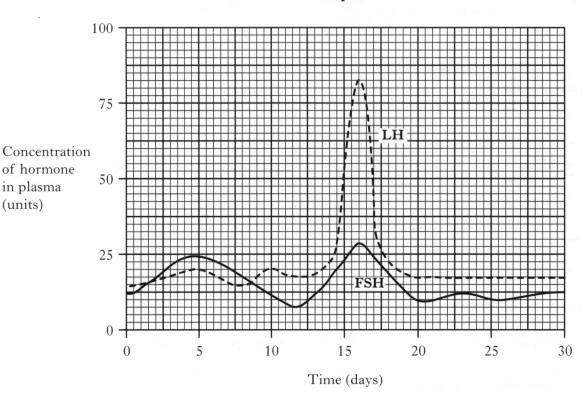

Graph 2

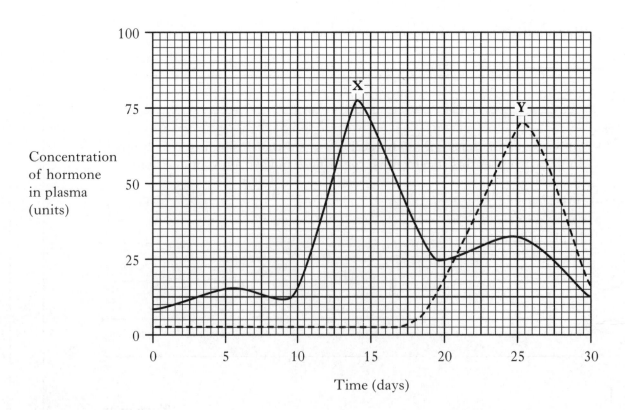

DO NOT WRITE IN THIS MARGIN

4. (continued) *Marks*

(*a*) Where in the body are FSH and LH produced?

____ **1**

(*b*) Name hormones X and Y.

X ____________________

Y ____________________ **1**

(*c*) What is the maximum concentration of hormone Y?

________ units **1**

(*d*) On which day did ovulation occur? Give a reason for your answer.

Day ________ **1**

Reason ____________________________________

____ **1**

(*e*) During her next cycle, the woman became pregnant.

Describe any differences which would occur in the concentrations of FSH and hormone Y after day 25.

FSH ____________________________________

____ **1**

Hormone Y ________________________________

____ **1**

[Turn over

Marks

5. (a) The table shows average quantities of substances filtered and excreted by the kidney per day.

Substance	Quantity filtered per day	Quantity excreted per day	Quantity reabsorbed per day
Water	$180 \, dm^3$	$1 \cdot 5 \, dm^3$	
Glucose	$175 \, g$	$0 \, g$	
Urea	$48 \, g$	$31 \, g$	
Protein	$0 \, g$	$0 \, g$	$0 \, g$

(i) Complete the table by calculating the quantities reabsorbed per day for water, glucose and urea.

1

(ii) What percentage of water filtered by the kidney is reabsorbed?

Space for calculation

_____ %

1

(iii) In which part of the kidney tubule is glucose reabsorbed?

_____

1

(b) Nephrosis is a kidney condition in which glomeruli are damaged.

As a result of nephrosis, the quantity of soluble proteins in the blood decreases and there is a build up of tissue fluid in the body.

(i) Explain why damage to the glomeruli results in a decrease of soluble protein in the blood.

_____

_____

1

(ii) Suggest a reason for the build-up of tissue fluid in the body.

_____

_____

1

Marks

6. The graph shows average blood pressure in different types of blood vessels.

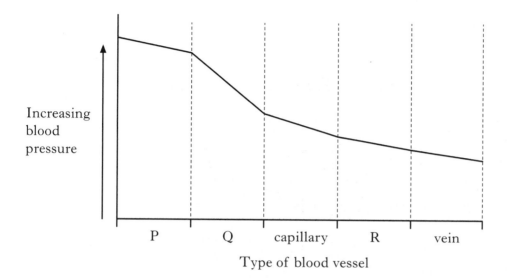

Type of blood vessel

(a) Name the types of blood vessel represented by P, Q and R.

P _____

Q _____

R _____   2

(b) Blood pressure values fluctuate in vessel type P.

Explain the reason for this.

_____

_____   1

(c) Explain why there is a large drop in blood pressure in vessel type Q.

_____

_____

_____   1

(d) In the vena cava, blood pressure falls below atmospheric air pressure yet blood is still able to return to the heart.

Explain how the blood flow is maintained.

_____

_____

_____

_____   2

DO N
WRIT
THI
MARC

7. An investigation was carried out to find out how a cyclist's metabolism changed while he pedalled at increasing speed.

Marks

The cyclist's heart rate, fat and carbohydrate consumption were measured at different power outputs.

The graph below shows the results of the investigation.

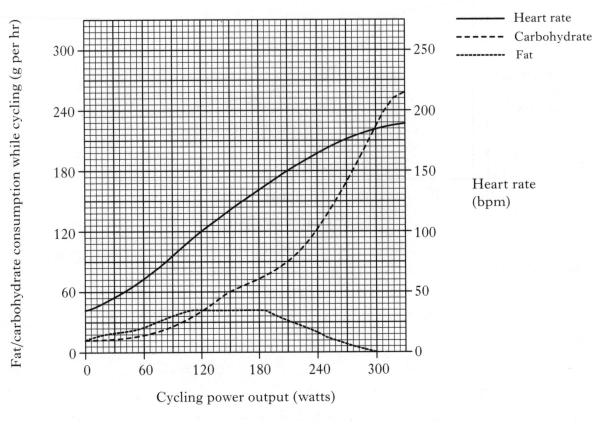

(a) What is the heart rate of the cyclist when his power output is 90 watts?

_____ bpm   1

(b) What evidence is there from the graph that the cyclist is very fit?

_____

_____   1

(c) Compare the consumption of fat and carbohydrate as cycling power increases. Quote data from the graph in your answer.

_____

_____

_____

_____

_____

_____

_____   3

Marks

7. **(continued)**

   (*d*) (i) Cyclists often use heart-rate monitors in training. A cyclist wishes to maintain his fat consumption at its maximum and, at the same time, limit his carbohydrate consumption.

   At what heart rate should he cycle?

   _____ bpm   1

   (ii) Suggest why it is good practice in a long distance cycling race to maximise fat consumption and minimise carbohydrate consumption.

   _____

   _____

   _____   1

   (*e*) The cyclist raced for 4 hours at a power output of 210 watts. During that time he consumed 100 g of carbohydrate in a liquid drink. Assuming he started with a carbohydrate store of 500 g, how much carbohydrate would he be left with at the end of the race?

   Space for calculation

   _____ g   1

   (*f*) (i) Glycogen is a major source of carbohydrate. Where is glycogen stored in the body?

   _____   1

   (ii) Name a hormone which promotes the conversion of glycogen to glucose.

   _____   1

   (iii) What substance is used as a source of energy after glycogen and fat stores have been used up?

   _____   1

   [Turn over

Marks

8. The diagrams below show two possible ways of classifying the nervous system.

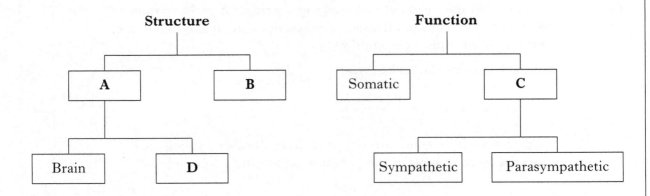

(a) (i) Identify A to D.

A _____

B _____

C _____

D _____   2

(ii) Describe **one** function of the somatic nervous system.

_____

_____

_____   1

(b) The brain contains two cerebral hemispheres.

(i) Name the structure which links these two hemispheres.

_____   1

(ii) The surfaces of the hemispheres are heavily folded to provide a large surface area.

Explain the significance of this feature.

_____

_____

_____   1

Marks

8.   (continued)

(*c*)   The diagram below shows some of the nerve connections between the brain
and three parts of the body.

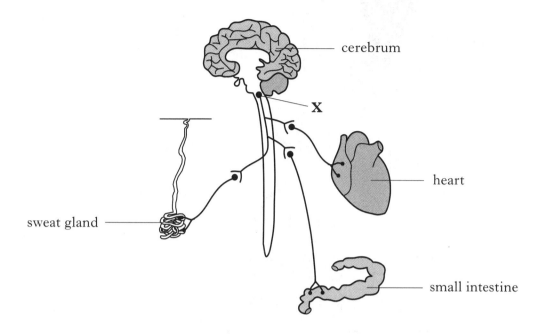

(i)   Identify the part of the brain labelled **X**.

_____   1

(ii)   The sympathetic and parasympathetic systems are often described as
antagonistic to one another.

Explain the meaning of *antagonistic*.

_____

_____   1

(iii)   Complete the table to show the effect of sympathetic stimulation on the
heart, sweat glands and small intestine.

Part of body	Sympathetic effect
Heart	
Sweat glands	
Small intestine	

2

[Turn over

Marks

DO NO
WRITE
THIS
MARG

9. The diagram shows how a non-biodegradable insecticide passes through a food chain in a Scottish fresh-water loch.

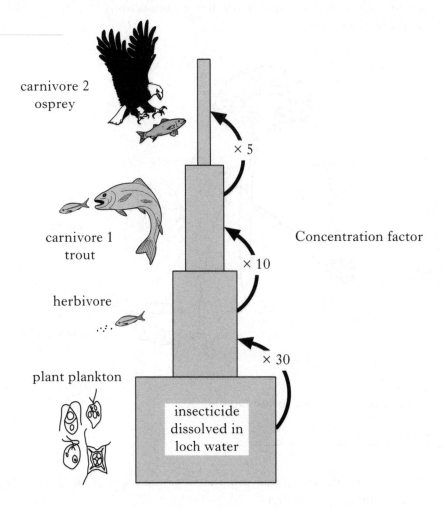

carnivore 2
osprey

× 5

carnivore 1
trout

Concentration factor

× 10

herbivore

plant plankton

× 30

insecticide
dissolved in
loch water

(a) Describe **one** way in which the insecticide could get into the loch water.

_____

_____

_____   1

(b) (i) The diagram shows the number of times the insecticide becomes concentrated at each stage of the food chain.

If the concentration of insecticide in the plant plankton is 0·025 ppm what would be the expected concentration in the osprey?

Space for calculation

_____ ppm   1

Marks

9. **(b)** **(continued)**

(ii) Explain why insecticide becomes more concentrated in carnivores at the top of the food chain.

_____

_____

_____

_____   **2**

(c) DDT is an insecticide which breaks down slowly at a rate of 50% every fifteen years. Calculate how long it would take for 100 kg of DDT to break down to less than 1 kg.

Space for calculation

_____ years   **1**

(d) Insecticides are chemicals used extensively in agriculture.

Name **two** other types of chemical used to treat crops and explain why they are used.

Chemical 1 _____

Use _____

_____   **1**

Chemical 2 _____

Use _____

_____   **1**

(e) Some insecticides work by disrupting enzyme-catalysed pathways.

What term is used to describe their action on enzymes?

_____   **1**

[Turn over

DO NO
WRITE
THIS
MARG

10. An experiment was carried out to investigate the effect of pH on the activity of the enzyme pepsin.

Marks

Six beakers were filled with pepsin solution and the pH adjusted in each beaker to give a range from pH 1 to pH 9. Six glass tubes were filled with egg albumen and boiled in water to set the egg white. The starting lengths of the egg white were measured and recorded in the table below.

The glass tubes were placed in the pepsin solution for a number of hours to allow digestion of the egg white. The lengths of egg white left in each tube at the end of the investigation are shown in the diagram below.

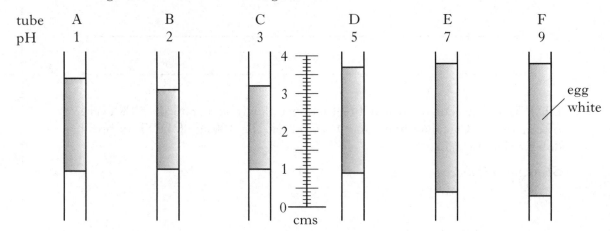

Tube	pH of pepsin solution	Length of egg white at start (mm)	Length of egg white at finish (mm)	Percentage decrease in length (%)
A	1	36	24	33
B	2	35	20	43
C	3	36		
D	5	34		
E	7	36	34	6
F	9	35	35	0

(a) (i) Complete the table above by measuring and recording the final lengths of egg white in tubes C and D.

1

(ii) Calculate the percentage decrease in length of egg white in tubes C and D and complete the table.

1

(b) Draw a line graph to show the relationship between pH and percentage decrease in length of egg white.

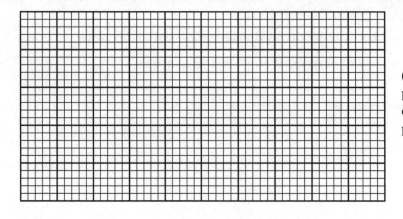

(Additional graph paper, if required, can be found on page 28.)

2

DO NOT
WRITE IN
THIS
MARGIN

Marks

10. **(continued)**

(c) (i) What conclusion can be drawn from the results of this experiment?

_____

_____ 1

(ii) Predict the percentage decrease in length of egg white in a pepsin solution of pH 12.

_____ 1

(iii) Why was it necessary to calculate a *percentage* decrease?

_____

_____ 1

(iv) Describe a suitable control for tube **A** in this investigation.

_____

_____ 1

(v) State **three** variables which would have to be kept constant throughout this investigation.

1 _____

2 _____

3 _____ 2

(vi) Describe **one** way in which the results could be made more reliable.

_____

_____ 1

(d) Pepsin is produced in an inactive form by cells lining the stomach.

Why is it important that pepsin is inactive when it is produced?

_____

_____

_____ 1

[Turn over

Marks

SECTION C

Both questions in this section should be attempted.

Note that each question contains a choice.

Questions 1 and 2 should be attempted on the blank pages which follow.

Supplementary sheets, if required, may be obtained from the invigilator.

Labelled diagrams may be used where appropriate.

1. Answer **either** A **or** B.

   A. Give an account of temperature regulation in cold conditions under the following headings:

       (i)   voluntary responses;   **3**

       (ii) involuntary responses;   **5**

       (iii) hypothermia.   **2**

             (10)

   OR

   B. Give an account of the development of boys at puberty under the following headings:

       (i)   physical changes;   **3**

       (ii) hormonal changes.   **7**

             (10)

In question 2, ONE mark is available for coherence and ONE mark is available for relevance.

2. Answer **either** A **or** B.

   A. Discuss how the impact of disease on the human population can be reduced.   **(10)**

   OR

   B. Describe the factors which influence the development of behaviour.   **(10)**

[END OF QUESTION PAPER]

DO NO
WRITE
THIS
MARG

SPACE FOR ANSWERS

SPACE FOR ANSWERS

SPACE FOR ANSWERS

DO N
WRITI
THI
MARG

SPACE FOR ANSWERS

ADDITIONAL GRAPH PAPER FOR QUESTION 10(b)

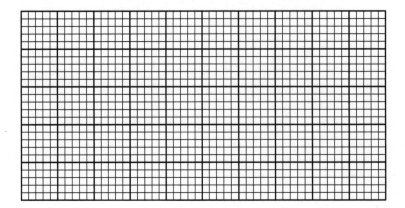

HIGHER

2009

[BLANK PAGE]

FOR OFFICIAL USE

Total for
Sections B & C

X009/301

NATIONAL
QUALIFICATIONS
2009

THURSDAY, 28 MAY
1.00 PM – 3.30 PM

HUMAN BIOLOGY
HIGHER

Fill in these boxes and read what is printed below.

Full name of centre

Town

Forename(s)

Surname

Date of birth

Day Month Year   Scottish candidate number   Number of seat

SECTION A–Questions 1—30

Instructions for completion of Section A are given on page two.

For this section of the examination you must use an **HB pencil**.

SECTIONS B AND C

1   (a) All questions should be attempted.

     (b) It should be noted that in **Section C** questions 1 and 2 each contain a choice.

2   The questions may be answered in any order but all answers are to be written in the spaces provided in this answer book, **and must be written clearly and legibly in ink**.

3   Additional space for answers will be found at the end of the book. If further space is required, supplementary sheets may be obtained from the invigilator and should be inserted inside the **front** cover of this book.

4   The numbers of questions must be clearly inserted with any answers written in the additional space.

5   Rough work, if any should be necessary, should be written in this book and then scored through when the fair copy has been written. If further space is required a supplementary sheet for rough work may be obtained from the invigilator.

6   Before leaving the examination room you must give this book to the invigilator. If you do not, you may lose all the marks for this paper.

Read carefully

1   Check that the answer sheet provided is for **Human Biology Higher (Section A)**.

2   For this section of the examination you must use an **HB pencil**, and where necessary, an eraser.

3   Check that the answer sheet you have been given has **your name**, **date of birth**, **SCN** (Scottish Candidate Number) and **Centre Name** printed on it.

    Do not change any of these details.

4   If any of this information is wrong, tell the Invigilator immediately.

5   If this information is correct, **print** your name and seat number in the boxes provided.

6   The answer to each question is **either** A, B, C or D. Decide what your answer is, then, using your pencil, put a horizontal line in the space provided (see sample question below).

7   There is **only one correct** answer to each question.

8   Any rough working should be done on the question paper or the rough working sheet, **not** on your answer sheet.

9   At the end of the exam, put the **answer sheet for Section A inside the front cover of this answer book**.

Sample Question

The digestive enzyme pepsin is most active in the

A   stomach

B   mouth

C   duodenum

D   pancreas.

The correct answer is **A**—stomach. The answer **A** has been clearly marked in **pencil** with a horizontal line (see below).

Changing an answer

If you decide to change your answer, carefully erase your first answer and, using your pencil, fill in the answer you want. The answer below has been changed to **D**.

SECTION A

All questions in this section should be attempted.

Answers should be given on the separate answer sheet provided.

1. Which of the following often act as a co-enzyme?

   A   Lipids

   B   Polysaccharides

   C   Hormones

   D   Vitamins

2. The table below refers to the mass of DNA in certain human body cells.

Cell type	Mass of DNA in cell ($\times 10^{-12}$ g)
liver	6·6
lung	6·6
P	3·3
Q	0·0

   Which of the following is most likely to identify correctly cell types P and Q?

	P	Q
A	kidney cell	sperm cell
B	sperm cell	mature red blood cell
C	mature red blood cell	sperm cell
D	nerve cell	mature red blood cell

3. The diagram below shows energy transfer within a cell.

   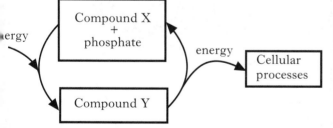

   Which line in the table below identifies correctly compounds X and Y?

	X	Y
A	glucose	ATP
B	glucose	ADP
C	ADP	ATP
D	ATP	glucose

4. The following chart shows stages in the complete breakdown of glucose in aerobic respiration.

   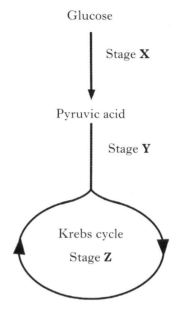

   At which stage or stages is hydrogen released to be picked up by hydrogen acceptors?

   A   Stages X, Y and Z

   B   Stages X and Y only

   C   Stages Y and Z only

   D   Stage Z only

5. The cell organelle shown below is magnified ten thousand times.

   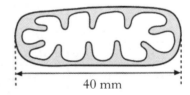

   What is the actual size of the organelle?

   A   0·04 μm

   B   0·4 μm

   C   4 μm

   D   40 μm

6. Lysosomes are abundant in

   A   enzyme secreting cells

   B   muscle cells

   C   cells involved in protein synthesis

   D   phagocytic cells.

7. The family tree below shows the transmission of the Rhesus D-antigen through three generations of a family. The allele coding for the presence of the Rhesus D-antigen is dominant and autosomal.

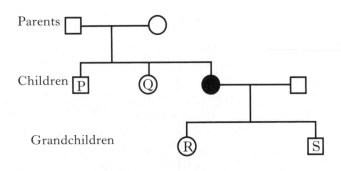

Parents

Children P   Q

Grandchildren R   S

☐ Rhesus positive male

■ Rhesus negative male

○ Rhesus positive female

● Rhesus negative female

Which of the children and grandchildren in the family tree must be heterozygous?

A   P, Q, R and S

B   P and Q only

C   R and S only

D   Q and R only

8. The table below shows some genotypes and phenotypes associated with a form of anaemia.

Genotype	Phenotype
AA	Unaffected
AS	Sickle cell trait
SS	Acute sickle cell anaemia

An unaffected person and someone with sickle cell trait have a child together.

What are the chances of the child having acute sickle cell anaemia?

A   none

B   1 in 4

C   1 in 2

D   1 in 1

9. The graph below shows changes which occur in the masses of protein, fat and carbohydrate in a person's body during seven weeks without food.

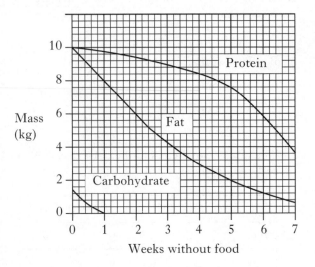

The person's starting weight was 60 kg. Predict their weight after two weeks without food.

A   57 kg

B   54 kg

C   50 kg

D   43 kg

10. The diagram below shows a section through seminiferous tubules in a testis.

Which cell produces testosterone?

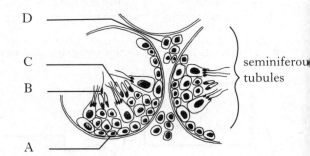

11. The diagram below represents part of the mechanism which controls ovulation.

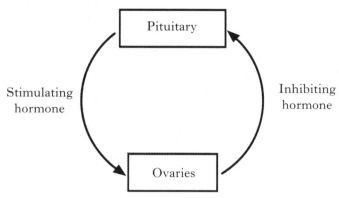

The hormones indicated above are

	Stimulating hormone	Inhibiting hormone
A	FSH	oestrogen
B	progesterone	FSH
C	oestrogen	LH
D	LH	testosterone

12. On which day in the following menstrual cycle could fertilisation occur?

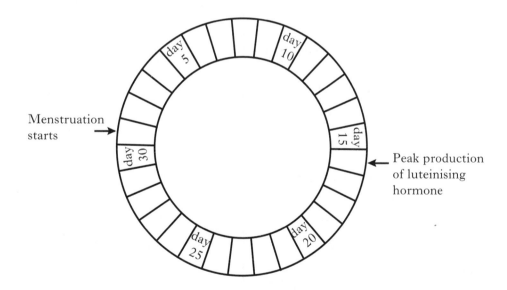

A   Day 30

B   Day 17

C   Day 14

D   Day 2

[Turn over

13. The diagram below shows blood vessels associated with the liver. The arrows show the direction of blood flow.

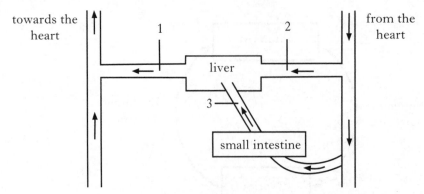

Which of the following correctly identifies the blood vessels.

	1	2	3
A	hepatic artery	hepatic vein	hepatic portal vein
B	hepatic vein	hepatic portal vein	hepatic artery
C	hepatic vein	hepatic artery	hepatic portal vein
D	hepatic artery	hepatic portal vein	hepatic vein

14. The relatively high urea concentration in the hepatic vein is a result of

   A   reabsorption of amino acids in the kidney

   B   conversion of glycogen to glucose in the liver

   C   deamination of amino acids in the liver

   D   excretion of amino acids in the kidney.

15. A person produces 0·75 litres of urine in 24 hours. The urine contains 18 g of urea.

What is the concentration of urea in the urine?

   A $1{\cdot}0\,g/100\,cm^3$

   B $1{\cdot}8\,g/100\,cm^3$

   C $2{\cdot}4\,g/litre$

   D $2{\cdot}4\,g/100\,cm^3$

16. The diagram below represents a part of the circulatory system of the skin.

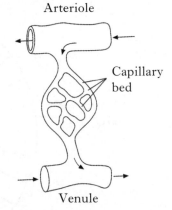

Which line in the table below correctly identifies changes which would take place in the blood as it flows from arteriole to venule?

	Concentration of	
	glucose	CO_2
A	increase	decrease
B	decrease	decrease
C	increase	increase
D	decrease	increase

17. A man was asked to breathe steadily at rest, then to breathe in and out as deeply as possible and finally to breathe steadily when exercising.

A trace of his lung capacity during this activity is shown.

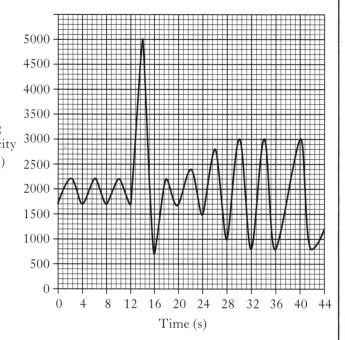

Time (s)

His volume of one breath at rest is

A   500 cm^3

B   2200 cm^3

C   4300 cm^3

D   5000 cm^3.

18. Which of the following is **not** a function of the lymphatic system?

A   It returns excess tissue fluid to the blood.

B   It causes the clotting of blood at wounds.

C   It destroys bacteria.

D   It transports fat from the small intestine.

19. When there is a decrease in the water concentration of the blood, which of the following series of events occur during the negative feedback response of the body?

	Concentration of ADH	Permeability of kidney tubules	Volume of urine
A	increases	increases	increases
B	decreases	decreases	increases
C	increases	increases	decreases
D	decreases	increases	decreases

20. Which of the following shows the correct responses to changes in blood sugar concentration?

	Sugar concentration in blood	Glucagon secretion	Insulin secretion	Glycogen stored in liver
A	increases	decreases	increases	increases
B	increases	decreases	increases	decreases
C	decreases	increases	decreases	increases
D	decreases	decreases	increases	decreases

[Turn over

21. High levels of blood glucose can cause clouding of the lens in the human eye. Concentrations above 5·5 mM are believed to put the individual at a high risk of lens damage.

In an investigation, subjects of different ages each drank a glucose solution. The concentration of glucose in their blood was then monitored over a number of hours. The results are shown in the graph below.

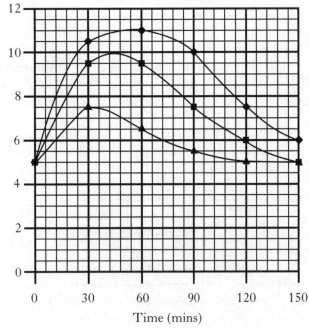

For how long during the investigation did 20 year olds remain above the high risk blood glucose concentration?

A   84 mins

B   90 mins

C   120 mins

D   148 mins

22. Which of the following parts of the brain is important in transferring information between the two cerebral hemispheres?

A   Hypothalamus

B   Corpus callosum

C   Cerebellum

D   Medulla oblongata

23. Which parts of the body are controlled by the largest motor area of the cerebrum?

A   Hands and lips

B   Feet and hands

C   Arms and hands

D   Legs and arms

24. The diagram below shows the ages in months at which children are able to walk unaided. The left end of the bar shows the age at which 25% of infants can walk unaided. The right end of the bar shows the age at which 90% of infants can walk unaided. The vertical bar shows the age at which 50% of infants can walk unaided.

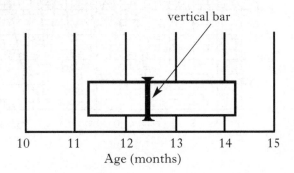

If sixteen infants, aged twelve months, were tested, how many would be expected to walk unaided?

A   4

B   7

C   9

D   12

25. Which of the following best describes memory span?

A   The total memory capacity of the brain

B   The time taken to learn a piece of information

C   The storage capacity of the short-term memory

D   The capacity to store information in long-term memory

26. The graph below shows the results of a survey carried out on members of an athletic club who ran an 800 m course under different conditions.

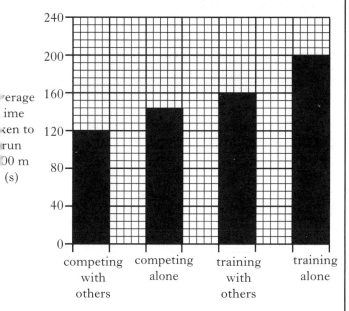

What is the percentage improvement in the time taken to run 800 m between those athletes training on their own and those training with others?

A   40%

B   25%

C   24%

D   20%

27. Which of the following processes reduces atmospheric carbon dioxide concentrations?

A   Decomposition

B   Nitrogen fixation

C   Respiration

D   Photosynthesis

28. Which of the following is a major source of methane?

A   Motor vehicles

B   Aerosols

C   Cattle

D   Nitrate fertilisers

29. The diagram below shows a nitrogen cycle associated with soil.

Which arrow indicates the activity of denitrifying bacteria?

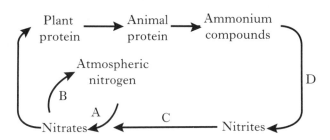

30. The age structure for four different human populations is represented in the diagrams below. The bars indicate the relative numbers in each age group.

Which diagram shows the population with the greatest potential for growth in the next forty years?

A   Age (years)

45–90

14–44

0–13

number in population

B   Age (years)

45–90

14–44

0–13

number in population

C   Age (years)

45–90

14–44

0–13

number in population

D   Age (years)

45–90

14–44

0–13

number in population

Candidates are reminded that the answer sheet MUST be returned INSIDE the front cover of this answer booklet.

DO NOT
WRITE
THIS
MARGI

SECTION B

Marks

All questions in this section should be attempted.

All answers must be written clearly and legibly in ink.

1. (a) The diagram below shows a section of a messenger RNA (mRNA) molecule.

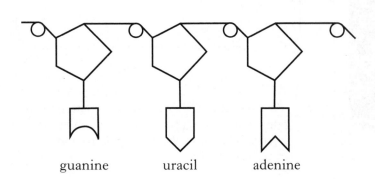

phosphate = O

sugar =

guanine   uracil   adenine

(i) Name the sugar that is present in mRNA.

_____   1

(ii) Which base found in mRNA is **not** shown in the diagram?

_____   1

(iii) Name **two** parts of a cell where mRNA is found.

1 _____

2 _____   1

(b) DNA templates are used to produce mRNA molecules.

(i) Insert the names of the DNA bases which pair with the RNA bases shown in the table below.

DNA base	RNA base
	adenine
	uracil
	guanine

1

DO NOT WRITE IN THIS MARGIN

Marks

1. (*b*) (continued)

(ii) Apart from free RNA nucleotides and a DNA template, name **one** other molecule that is essential for mRNA synthesis.

______________________________ **1**

(iii) Describe the part played by an mRNA molecule in the manufacture of a cell protein.

____

____

____

____ **3**

[Turn over

Marks

2. Many inherited disorders are caused by inborn errors of metabolism.

    (*a*)   (i)   What causes disorders that lead to inborn errors of metabolism?

    _____   1

            (ii)   How do these inherited disorders affect metabolic pathways?

    _____

    _____   1

    (*b*)   Phenylketonuria (PKU) is an example of an inherited disorder.

    One metabolic pathway affected by PKU is shown below.

    | *enzyme 1* | *enzyme 2* | *enzyme 3* |

    phenylalanine ⟶ tyrosine ⟶ intermediate ⟶ noradrenaline
    compounds

            (i)   Describe how PKU affects the metabolic pathway shown above.

    _____

    _____   1

            (ii)   With reference to the metabolic pathway shown, explain why PKU affects the nervous system.

    _____

    _____

    _____

    _____   2

    (*c*)   What term describes the testing of newborn babies for inherited disorders such as PKU?

    _____   1

Marks

3. (*a*) The MN blood group system is determined by two alleles, M and N. Each of these alleles controls the production of a different antigen on the cell membrane of red blood cells.

M and N are co-dominant.

(i) Two parents, who are heterozygous for this blood group, have a son. Complete the Punnett square below to show the parental gametes and the possible genotypes of their son.

Parents   Mother   Father
Genotype   MN   ×   MN

		mother's gametes	
father's gametes			MN

1

(ii) The son has a different genotype to either of his parents. What are the chances of this happening?

Space for calculation

_____ %   **1**

(iii) Describe how the son's phenotype differs from his parents.

_____

_____   **1**

(*b*) The immune system recognises antigens on the cell membrane as self or non-self.

What term describes

(i) an immune reaction to self antigens?

_____   **1**

(ii) an over-reaction to a normally harmless non-self antigen?

_____   **1**

[Turn over

Marks

4. Hydrogen peroxide is a toxic chemical which is produced during metabolism. Catalase is an enzyme which breaks down hydrogen peroxide as shown below.

$$\text{hydrogen peroxide} \xrightarrow{\text{catalase}} \text{water} + \text{oxygen}$$

Experiments were carried out to investigate how changing the concentration of catalase affects the rate of hydrogen peroxide breakdown.

Filter paper discs were soaked in catalase solutions of different concentration. Each disc was then added to a beaker of hydrogen peroxide solution as shown in **Figure 1**.

The disc sank to the bottom of the beaker before rising back up to the surface. The time taken for each disc to rise to the surface was used to measure the reaction rates.

The results of the investigation are shown in **Table 1**.

Figure 1

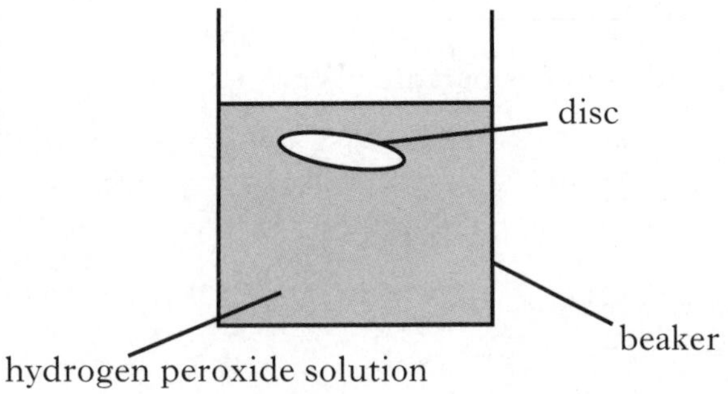

Table 1

catalase concentration (%)	average time for ten discs to rise (s)
0·125	9·8
0·25	6·9
0·5	5·0
1·0	3·8
2·0	3·8

(a) Explain why the filter paper discs rose to the surface of the hydrogen peroxide solution.

_____

_____   1

(b) Name **three** variables which should be controlled during this investigation.

1 _____

2 _____

3 _____   2

(c) What feature of **this** investigation makes the results more reliable?

_____

_____   1

Marks

4. (continued)

(d) It was suggested that the filter paper was reacting with the hydrogen peroxide. How could this be tested using the same procedure?

_____

_____   **1**

(e) (i) Plot a line graph to illustrate the results of the investigation. (Additional graph paper, if required, can be found on *Page thirty-two*.)

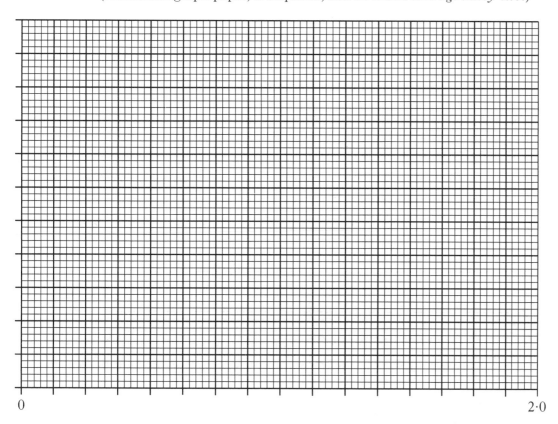

0   2·0   **2**

(ii) State **two** conclusions which can be drawn from these results.

1 _____

_____

2 _____

_____   **2**

(f) Explain why the addition of an inhibitor would slow down the rate of this reaction.

_____

_____   **1**

Marks

5. The diagram shows part of the reproductive system of a woman in early pregnancy.

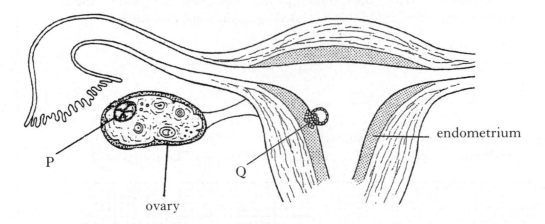

P

ovary

Q

endometrium

(a) Place an **X** on the diagram to show where fertilisation occurred.   1

(b) Structure P produces progesterone at this stage in pregnancy.

   (i) Name structure P.

   _____   1

   (ii) State **one** function of progesterone during early pregnancy.

   _____

   _____   1

(c) Structure Q will develop into the placenta.

   Name the processes involved in the transfer of oxygen, glucose and antibodies across the placenta.

   Oxygen _____

   Glucose _____

   Antibodies _____   2

(d) In the early stages of pregnancy the cells of the embryo are starting to differentiate.

   Describe what happens during differentiation.

   _____

   _____   1

DO NOT WRITE IN THIS MARGIN

Marks

5. (continued)

(*e*) Name a stage of embryo development that comes between fertilisation and differentiation.

____________________ **1**

(*f*) A woman gives birth to monozygotic twins.

State whether monozygotic twins are identical or non-identical and give a reason for your answer.

Monozygotic twins ____________________

Reason ____

____

____ **1**

[Turn over

DO NC
WRITE
THI
MARG

6. The diagram below shows stages in the life history of a red blood cell.   *Marks*

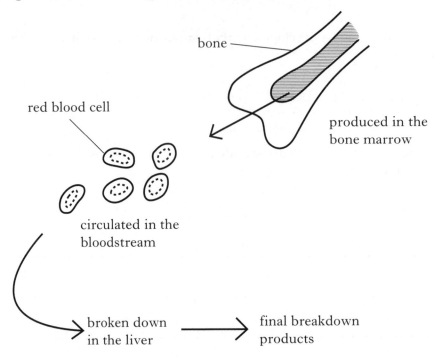

(a) Vitamin B_{12} and iron are both used in the production of red blood cells.

    (i) What substance is needed for the absorption of Vitamin B_{12} from the gut?

       _____   1

    (ii) Which molecule requires iron for its production?

       _____   1

(b) On average, how long do red blood cells remain in circulation?

    _____   1

(c) At any given time there are 5·5 million red blood cells in 1 millilitre of human blood.

    Calculate how many red blood cells will be in the circulation of an individual who has a total blood volume of 5 litres.

    Space for calculation

                                         _____ million   1

DO NOT
WRITE IN
THIS
MARGIN

Marks

6. **(continued)**

(*d*) Explain how the structure of a red blood cell

    (i)   makes it very efficient at absorbing oxygen.

    _____

    _____   1

    (ii) allows it to pass through capillaries.

    _____

    _____   1

(*e*) Apart from the liver, name a body site where red blood cells are broken down.

    _____   1

(*f*) One of the final products of the breakdown of red blood cells is bile.

    (i)   Where is bile stored in the body?

    _____   1

    (ii) Explain the importance of bile salts in the digestion of lipids.

    _____

    _____

    _____

    _____   2

[Turn over

DO NO
WRITE
THI
MARG

7.   Oxygen consumption is often used to measure the intensity of exercise.

VO_{2max} is the maximum rate at which someone can take up and use oxygen.   *Marks*

Graph 1 shows the VO_{2max} of office workers, and various professional sportsmen and sportswomen.

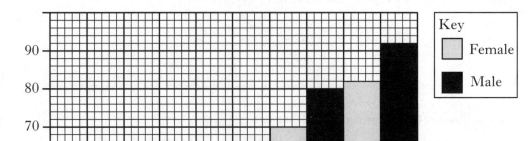

(a)   (i)   What is the difference between the VO_{2max} of a male cross-country skier and a male office worker?

Space for calculation

_____   1

(ii)   Cross-country skiing is a very energy demanding sport.

What is the advantage to a cross-country skier of having a high VO_{2max}?

_____

_____   1

(b)   Calculate the oxygen uptake, during a three minute race, of a female rower who weighs 85 kg. Assume that she has maximum oxygen uptake throughout the race.

Space for calculation

_____ litres   1

(c)   The graph shows that, on average, men have higher maximum oxygen uptakes than women.

Suggest a reason for this difference.

_____

_____   1

Marks

7. (continued)

Tests which determine the VO_{2max} of individuals use the relationship between heart rate and oxygen uptake.

The maximum oxygen uptake occurs when an individual's heart is beating at its maximum rate.

Graph 2 shows measurements of heart rate and oxygen uptake for a professional sportsman and an office worker, who are both 24 years old. The measurements were taken as speed was gradually increased on a treadmill.

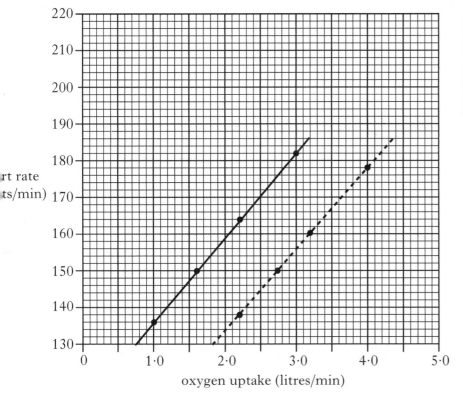

heart rate
(beats/min)

oxygen uptake (litres/min)

Key

——— Office worker

- - - - - Sportsman

(*d*) (i) An individual's maximum heart rate can be calculated by subtracting their age from 220.

Calculate the maximum heart rate of the office worker.

Space for calculation

_____ beats/min   **1**

(ii) **Use the graph** to predict the maximum oxygen uptake of the office worker.

_____ litres/min   **1**

(iii) The sportsman weighed 60 kg.

Use the information in **graphs 1** and **2** to determine his sport.

_____   **1**

Marks

8. The diagram below shows the human heart and some associated blood vessels. The arrows on the diagram show the direction of blood flow.

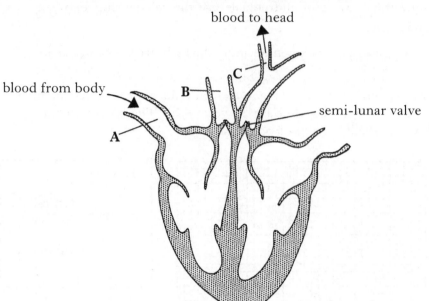

blood to head

blood from body

B

C

A

semi-lunar valve

(a) Name blood vessels **A**, **B** and **C**.

A _____

B _____

C _____   2

(b) Place arrows on the diagram to show the path of oxygenated blood as it flows through the heart.   1

(c) Describe the function of the semi-lunar valve **labelled on the diagram**.

_____

_____   1

(d) During which stage of the cardiac cycle do the semi-lunar valves open?

_____   1

Marks

9. (a) The diagram below shows a synapse which links a nerve cell with the sinoatrial node (SAN) in the heart.

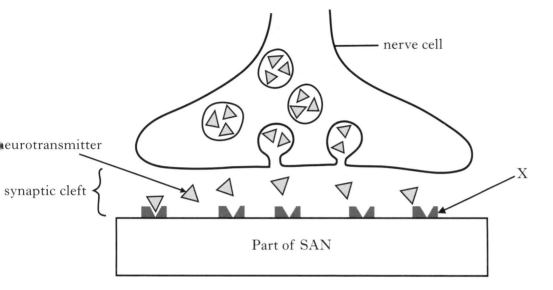

nerve cell

neurotransmitter

synaptic cleft

X

Part of SAN

(i) Where in the heart is the SAN located?

_____   1

(ii) Describe the function of molecule X.

_____   1

(b) One example of a neurotransmitter is acetylcholine.

How is acetylcholine removed from the synapse?

_____

_____   1

(c) (i) In which area of the brain does the sympathetic nervous system originate?

_____   1

(ii) Describe a situation which would lead to stimulation of the sympathetic nervous system.

_____

_____   1

DO NO
WRITE
THIS
MARG

10. The diagram below shows two different neural pathways.

Nerve impulses are travelling from left to right in both pathways.

Marks

Pathway A

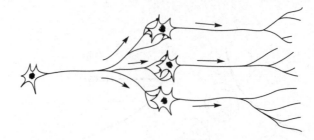

Pathway B

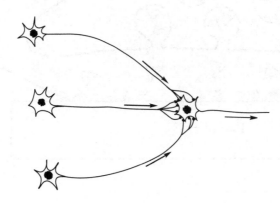

(a) (i) Name the types of pathway represented by **A** and **B**.

A _____

B _____   1

(ii) Pathway **A** helps the hand to function.

Explain how it does this.

_____

_____

_____   2

(b) Blinking is a reflex action.

(i) What is a reflex action?

_____

_____   1

(ii) The blinking reflex can sometimes be suppressed.

What term refers to the ability of the nervous system to suppress reflexes?

_____   1

11. One-fifth of all UK deaths are caused by smoking.

Marks

The graph below shows the total number of deaths from lung cancer of males and females of different ages in the United Kingdom in 2004.

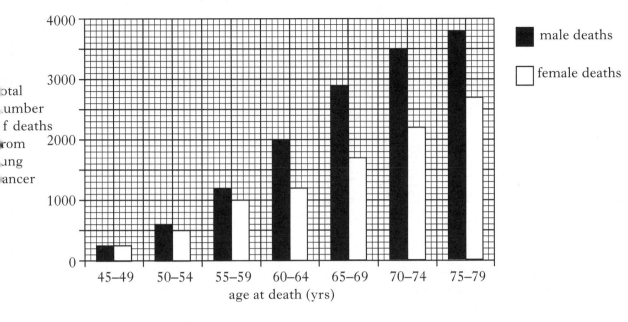

Total number of deaths from lung cancer

age at death (yrs)

■ male deaths

□ female deaths

(*a*) Describe the **two** main trends shown by the graph.

1 _____

_____

2 _____

_____   **2**

(*b*) (i) Calculate the whole number ratio of male to female deaths in 45 to 49 year olds and 60 to 64 year olds.

Space for calculation

45–49 years _____ : _____   60–64 years _____ : _____
                male   female   male   female   **1**

(ii) Suggest a reason for the **difference** between the two calculated ratios.

_____

_____   **1**

(*c*) Ninety-five percent of deaths from lung cancer occur in smokers.

Calculate how many male non-smokers aged 75 to 79 died from lung cancer in the UK in 2004.

Space for calculation

_____   **1**

DO NO
WRITE
THIS
MARG

Marks

12. The map below represents a short length of a Scottish river.

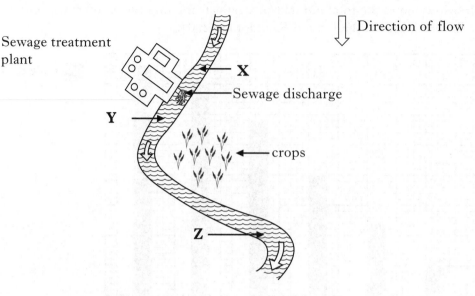

Sewage treatment plant

⇩ Direction of flow

X

Sewage discharge

Y →

← crops

Z →

(a) The sewage treatment works sometimes gets overloaded and untreated sewage is discharged into the river.

   (i) Following the discharge of sewage, state how bacteria would change in number between the following points.

     Give a reason for your answer.

     A Between points **X** and **Y**.

       Change _____

       Reason _____

       _____ 1

     B Between points **Y** and **Z**.

       Change _____

       Reason _____

       _____ 1

   (ii) State how algae would change in number between points **Y** and **Z**.

     Give a reason for your answer.

     Change _____

     Reason _____

     _____ 1

DO NOT WRITE IN THIS MARGIN

12. (continued)

Marks

(*b*) Herbicides are frequently applied to land where crops are growing.

(i) What is a herbicide?

______________________________ **1**

(ii) Explain how the use of herbicides leads to an increased crop yield.

____

______________________________ **1**

(*c*) Crop yield can be increased by the insertion of a gene from another organism into a chromosome of the crop plant.

Name this process.

______________________________ **1**

[Turn over for Section C

Marks

SECTION C

Both questions in this section should be attempted.

Note that each question contains a choice.

Questions 1 and 2 should be attempted on the blank pages which follow.

Supplementary sheets, if required, may be obtained from the invigilator.

Labelled diagrams may be used where appropriate.

1. Answer **either** A **or** B.

    A. Discuss how other people can affect an individual's behaviour under the following headings:

        (i) the influence of groups;   **6**

        (ii) influences that change beliefs.   **4**

        (10)

    OR

    B. Discuss global warming under the following headings:

        (i) possible causes of global warming;   **6**

        (ii) potential effects of rising sea levels.   **4**

        (10)

In question 2, ONE mark is available for coherence and ONE mark is available for relevance.

2. Answer **either** A **or** B.

    A. Describe how immunity is naturally acquired.   **(10)**

    OR

    B. Describe the nature and reproduction of viruses.   **(10)**

[END OF QUESTION PAPER]

SPACE FOR ANSWERS

SPACE FOR ANSWERS

SPACE FOR ANSWERS

ADDITIONAL GRAPH FOR QUESTION 4(*e*)

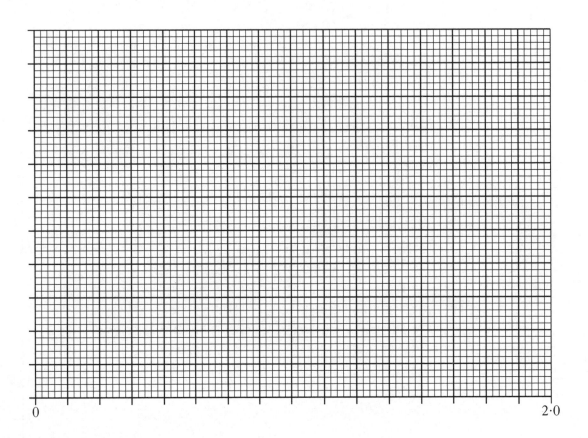

0   2·0

[BLANK PAGE]

FOR OFFICIAL USE

Total for
Sections B & C

X009/301

NATIONAL
QUALIFICATIONS
2010

THURSDAY, 27 MAY
1.00 PM – 3.30 PM

HUMAN BIOLOGY
HIGHER

Fill in these boxes and read what is printed below.

Full name of centre

Town

Forename(s)

Surname

Date of birth

Day	Month	Year

Scottish candidate number

Number of seat

SECTION A—(30 marks)

Instructions for completion of Section A are given on page two.

For this section of the examination you must use an **HB pencil**.

SECTIONS B AND C—(100 marks)

1 (a) All questions should be attempted.

  (b) It should be noted that in **Section C** questions 1 and 2 each contain a choice.

2 The questions may be answered in any order but all answers are to be written in the spaces provided in this answer book, **and must be written clearly and legibly in ink**.

3 Additional space for answers will be found at the end of the book. If further space is required, supplementary sheets may be obtained from the Invigilator and should be inserted inside the **front** cover of this book.

4 The numbers of questions must be clearly inserted with any answers written in the additional space.

5 Rough work, if any should be necessary, should be written in this book and then scored through when the fair copy has been written. If further space is required a supplementary sheet for rough work may be obtained from the Invigilator.

6 Before leaving the examination room you must give this book to the Invigilator. If you do not, you may lose all the marks for this paper.

Read carefully

1   Check that the answer sheet provided is for **Human Biology Higher (Section A)**.

2   For this section of the examination you must use an **HB pencil**, and where necessary, an eraser.

3   Check that the answer sheet you have been given has **your name**, **date of birth**, **SCN** (Scottish Candidate Number) and **Centre Name** printed on it.

     Do not change any of these details.

4   If any of this information is wrong, tell the Invigilator immediately.

5   If this information is correct, **print** your name and seat number in the boxes provided.

6   The answer to each question is **either** A, B, C or D. Decide what your answer is, then, using your pencil, put a horizontal line in the space provided (see sample question below).

7   There is **only one correct** answer to each question.

8   Any rough working should be done on the question paper or the rough working sheet, **not** on your answer sheet.

9   At the end of the examination, put the **answer sheet for Section A inside the front cover of this answer book**.

Sample Question

The digestive enzyme pepsin is most active in the

A   stomach

B   mouth

C   duodenum

D   pancreas.

The correct answer is **A**—stomach. The answer **A** has been clearly marked in **pencil** with a horizontal line (see below).

A   B   C   D

Changing an answer

If you decide to change your answer, carefully erase your first answer and, using your pencil, fill in the answer you want. The answer below has been changed to **D**.

A   B   C   D

SECTION A

All questions in this section should be attempted.

Answers should be given on the separate answer sheet provided.

1. The diagram below shows an enzyme-catalysed reaction taking place in the presence of an inhibitor.

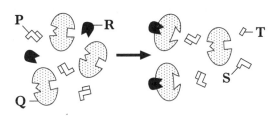

Which line in the table below identifies correctly the molecules in the reaction?

	Inhibitor	Substrate	Product
A	P	R	S
B	Q	P	S
C	R	P	T
D	R	Q	T

2. The following diagram shows a branched metabolic pathway.

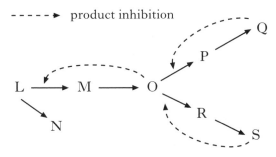

- - - - - ▶ product inhibition

Which reaction would tend to occur if both Q and S are present in the cell in high concentrations?

A   L ⟶ M

B   R ⟶ S

C   O ⟶ P

D   L ⟶ N

3. A fragment of DNA was found to have 120 guanine bases and 60 adenine bases. What is the total number of sugar molecules in this fragment?

A   60

B   90

C   180

D   360

4. The following information refers to protein synthesis.

tRNA anticodon	amino acid carried by tRNA
G U G	Histidine (his)
C G U	Alanine (ala)
G C A	Arginine (arg)
A U G	Tyrosine (tyr)
U A C	Methionine (met)
U G U	Threonine (thr)

What order of amino acids would be synthesised from the base sequence of DNA shown?

Base sequence of DNA

C   G   T   T   A   C   G   T   G

A   arg   -   tyr   -   his

B   ala   -   met   -   his

C   ala   -   tyr   -   his

D   arg   -   tyr   -   thr

5. In which of the following is the cell organelle listed correctly with its function?

	Cell organelle	Function
A	Mitochondrion	Anaerobic respiration
B	Ribosome	Release of ATP
C	Lysosome	Synthesis of enzymes
D	Nucleolus	Synthesis of RNA

6. Carrier molecules involved in the process of active transport are made of

A   protein

B   carbohydrate

C   lipid

D   phospholipid.

7. An investigation was carried out into the uptake of sodium ions by animal cells. The graph shows the rates of sodium ion uptake and breakdown of glucose at different concentrations of oxygen.

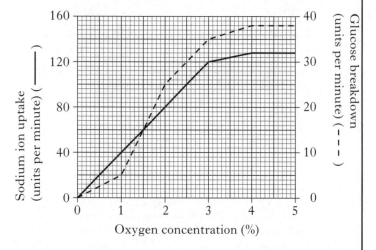

Calculate the number of units of sodium ions that are taken up over a 5 minute period when the concentration of oxygen in solution is 2%.

A   80

B   100

C   400

D   500

8. Which of the following statements about viruses is true?

A   Viral protein directs the synthesis of new viruses.

B   New viruses are assembled outside the host cell.

C   Viral protein is injected into the host cell.

D   Viral DNA directs the synthesis of new viruses.

9. What is the significance of chiasma formation?

A   It results in the halving of the chromosome number.

B   It results in the pairing of homologous chromosomes.

C   It permits gene exchange between homologous chromosomes.

D   It results in the independent assortment of chromosomes.

10. The transmission of a gene for deafness is shown in the family tree below.

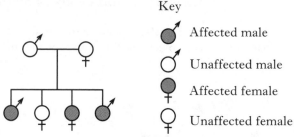

Key

Affected male

Unaffected male

Affected female

Unaffected female

This condition is controlled by an allele which is

A   dominant and sex-linked

B   recessive and sex-linked

C   dominant and not sex-linked

D   recessive and not sex-linked.

11. The examination of a karyotype would **not** detect

A   phenylketonuria

B   Down's syndrome

C   the sex of the fetus

D   evidence of non-disjunction.

12. A woman with blood group *AB* has a child to a man with blood group *O*. What are the possible phenotypes of the child?

A   *A* or *B*

B   *AB* only

C   *AB* or *O*

D   *AB, A* or *B*

13. Cystic fibrosis is an inherited condition caused by a recessive allele. The diagram below is a family tree showing affected individuals.

unaffected male
affected male
unaffected female
affected female

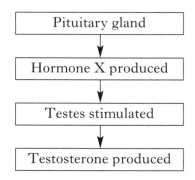

Which two individuals in this family tree must be heterozygous for the cystic fibrosis gene?

A   3 and 5

B   4 and 6

C   1 and 2

D   2 and 6

14. The diagram below shows the influence of the pituitary gland on testosterone production.

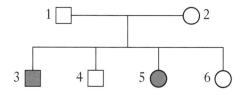

What is hormone X?

A   Luteinising hormone

B   Follicle stimulating hormone

C   Oestrogen

D   Progesterone

15. From which structure in the female reproductive system does a corpus luteum develop?

A   Endometrium

B   Graafian follicle

C   Fertilised ovum

D   Unfertilised ovum

16. The table below contains information about four semen samples.

| | Semen sample | | | |
	A	B	C	D
Number of sperm in sample (millions/cm³)	40	30	20	60
Active sperm (percent)	50	60	75	40
Abnormal sperm (percent)	30	65	10	70

Which semen sample has the highest number of active sperm per cm³?

17. Which of the following describes correctly the exchange of materials between maternal and fetal circulations?

	Glucose	Antibodies
A	into fetus by active transport	into fetus by active transport
B	into fetus by active transport	into fetus by pinocytosis
C	into fetus by pinocytosis	into fetus by active transport
D	into fetus by diffusion	into mother by pinocytosis

18. The diffusion pathway of carbon dioxide within body tissues is

A   plasma → tissue fluid → cell cytoplasm

B   lymph → tissue fluid → cell cytoplasm

C   cell cytoplasm → tissue fluid → plasma

D   tissue fluid → lymph → plasma.

[Turn over

19. The graph below shows changes in arterial blood pressure.

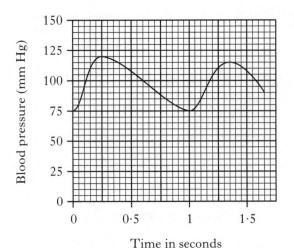

Time in seconds

The shape of the graph is due to

A   the action of the heart muscle

B   the action of the diaphragm

C   the closing of the valves in the veins

D   muscular contraction of the arteries.

20. An ECG trace is shown below.

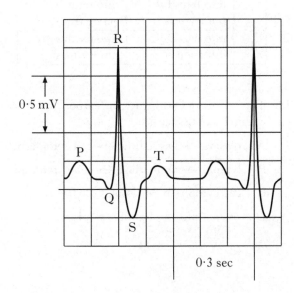

0·3 sec

What is the person's heart rate?

A   100 beats per minute

B   120 beats per minute

C   150 beats per minute

D   200 beats per minute

21. Which of the following statements refers correctly to the cardiac cycle?

A   During systole the atria contract followed by the ventricles.

B   During systole the ventricles contract followed by the atria.

C   During diastole the atria contract followed by the ventricles.

D   During diastole the ventricles contract followed by the atria.

22. Which line in the table below correctly describes the conditions under which the affinity of haemoglobin for oxygen is highest?

	Oxygen tension	Temperature (°C)
A	high	40
B	high	37
C	low	37
D	low	40

23. Which of the following is triggered by the hypothalamus in response to an increase in the temperature of the body?

A   Contraction of the hair erector muscles and vasodilation of the skin arterioles

B   Contraction of the hair erector muscles and vasoconstriction of the skin arterioles

C   Relaxation of the hair erector muscles and vasodilation of the skin arterioles

D   Relaxation of the hair erector muscles and vasoconstriction of the skin arterioles

24. The graph below shows the rate of sweating of an individual in different environmental conditions.

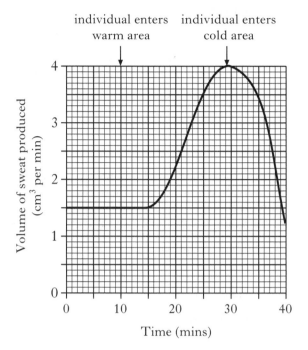

How long after entering the warm area did it take for the volume of sweat production to increase by 100%?

A   8 minutes

B   13 minutes

C   20 minutes

D   23 minutes

25. The diagram below shows the main parts of the brain as seen in vertical section.

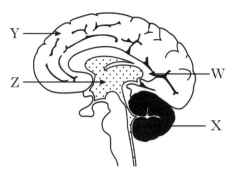

Which line in the table below correctly identifies the functions of two areas of the brain?

	Communication between hemispheres	Reasoning
A	W	X
B	X	Y
C	W	Y
D	Z	W

[Turn over

26. The diagram below shows a test on a man who had a damaged corpus callosum. This meant that he could no longer transfer information between his right and left cerebral hemispheres.

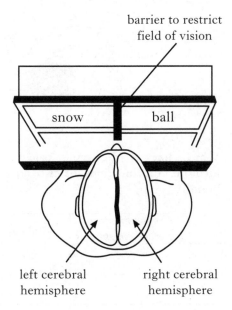

barrier to restrict field of vision

snow   ball

left cerebral hemisphere   right cerebral hemisphere

Some of the functions of each hemisphere are described in the table below.

Left cerebral hemisphere	Right cerebral hemisphere
processes information from right eye	processes information from left eye
controls language production	controls spatial task co-ordination

The man was asked to look straight ahead and then the words "snow" and "ball" were flashed briefly on the screen as shown.

What would the man say that he had just seen?

A   Ball

B   Snow

C   Snowball

D   Nothing

27. Which of the following statements about diverging neural pathways is correct?

A   They accelerate the transmission of sensory impulses.

B   They suppress the transmission of sensory impulses.

C   They decrease the degree of fine motor control.

D   They increase the degree of fine motor control.

28. Which of the following describes the change in an individual's behaviour where the presence of others causes the individual to show less restraint and become more impulsive?

A   Social facilitation

B   Shaping

C   Generalisation

D   Deindividuation

29. Which of the following identifies correctly a process in the nitrogen cycle?

A   Nitrifying bacteria trap atmospheric nitrogen.

B   Nitrifying bacteria convert ammonium compounds to nitrates.

C   Nitrogen-fixing bacteria convert nitrates to atmospheric nitrogen.

D   Denitrifying bacteria convert ammonia to nitrates.

30. The diagrams below contain information about the population of Britain.

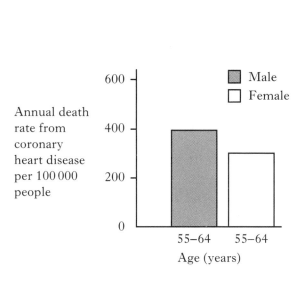

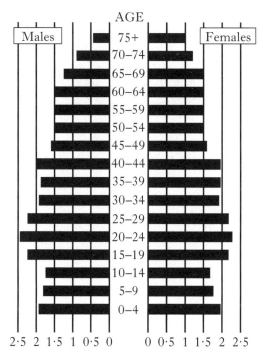

Population size (millions)

How many British men between 55 and 64 years of age die from coronary heart disease annually?

A   400

B   6000

C   12 000

D   24 000

Candidates are reminded that the answer sheet MUST be returned INSIDE the front cover of this answer booklet.

[Turn over for Section B on *Page eleven*

[BLANK PAGE]

Marks

SECTION B

All questions in this section should be attempted.

All answers must be written clearly and legibly in ink.

1.   The diagram below represents stages in the production of human sperm.

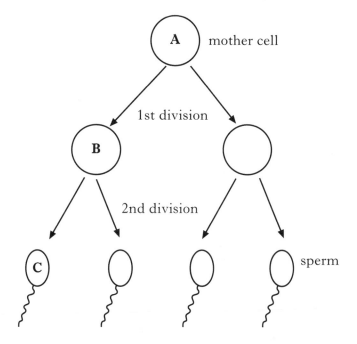

(a)   Name the type of cell division that produces sex cells.

_____   1

(b)   State the number of chromosomes which would be present in the cells labelled A, B and C.

A: _____   B: _____   C: _____   1

(c)   Compare the appearance of the chromosomes in cell B and cell C.

_____

_____   1

(d)   Name the **two** processes which increase variation during the 1st division of the sperm mother cell.

1 _____

2 _____   1

(e)   State the location of sperm production in the testes.

_____   1

Marks

2.   The diagram below shows some of the reactions which occur during aerobic respiration.

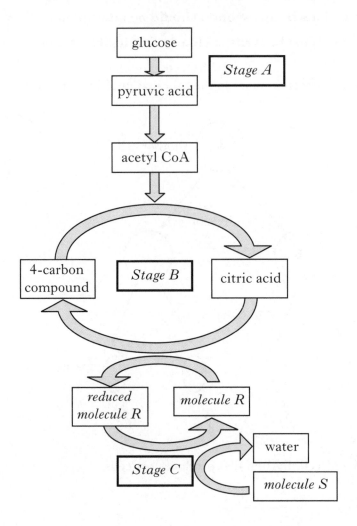

(a)   Complete the table by naming stages A, B and C and indicating their **exact** location within the cell.

Stage	Name	Location
A		
B		
C		

3

(b)   A glucose molecule contains 6 carbon atoms.

How many carbon atoms are found in the following molecules?

Pyruvic acid   _____

Citric acid   _____   1

DO NOT
WRITE IN
THIS
MARGIN

Marks

2. **(continued)**

(*c*) Complete the following sentences by naming molecules R and S and describing their function with respect to stage C.

R is _____ and its function is _____

_____

_____ .

S is _____ and its function is _____

_____

_____ .　　**2**

(*d*) Under normal circumstances carbohydrate is the main respiratory substrate.

In each of the following extreme situations, state the alternative respiratory substrate and explain why the body has to use it.

Situation	Respiratory substrate	Explanation
Prolonged starvation		
Towards the end of a marathon race		

2

[Turn over

Marks

3.   The diagram below shows blood from a person who has been infected by bacteria.
     These bacteria have triggered an immune response involving proteins P and Q.

The diagram is not drawn to scale.

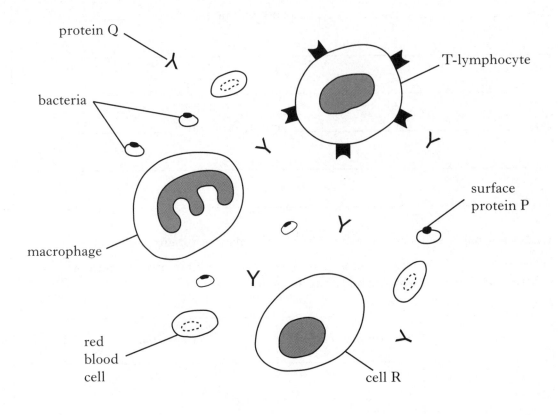

(a)   (i)   Identify proteins P and Q.

P _____   Q _____   1

      (ii)   Cell R produced protein Q.
             Name this type of cell.

             _____   1

      (iii) Describe the role of the following cells in combating infection.

             (A)   T-lymphocyte _____

                   _____   1

             (B)   Macrophage _____

                   _____   1

DO NOT WRITE IN THIS MARGIN

Marks

3. (continued)

(*b*) Complete the following sentences by underlining one option from each pair of options shown in **bold**.

(i) Immunity gained after contracting a bacterial infection is an example of **active** / **passive** immunity that is **naturally** / **artificially** acquired. **1**

(ii) Immunity gained from the injection of a tetanus vaccine is an example of **active** / **passive** immunity that is **naturally** / **artificially** acquired. **1**

(*c*) What happens during an autoimmune response?

____

____ **1**

[Turn over

Marks

4. Lactose is the main sugar found in milk.

    Lactose is broken down by lactase, an enzyme which is made by cells lining the small intestine. The glucose and galactose molecules produced are then absorbed into the bloodstream.

$$\text{lactose} \xrightarrow{\text{lactase}} \text{glucose} + \text{galactose}$$

A student carried out an investigation to compare the lactose content of human milk and cow milk.

He set up a test tube containing human milk and lactase solution. Every 30 seconds samples were taken and the glucose concentration measured. Then he repeated the procedure with cow milk.

His experimental setup is shown in Figure 1.

His results are shown in the table below.

Time (min)	Concentration of glucose (%)	
	Human milk	Cow milk
0	0	0
0·5	0·28	0·28
1·0	0·54	0·46
1·5	0·80	0·54
2·0	1·04	0·58
2·5	1·10	0·58
3·0	1·10	0·58

Figure 1

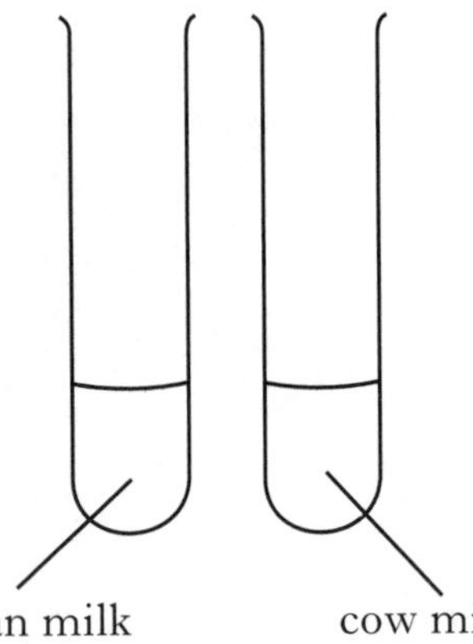

human milk
and
lactase

cow milk
and
lactase

(a) Lactose is a disaccharide sugar.

Explain how the information above supports this statement.

_____

_____   1

(b) One variable that must be kept constant in this investigation is pH.

List **two** other variables which would have to be kept constant.

1 _____

2 _____   1

Marks

4. (continued)

(c) Construct a line graph to show all the data in the table.

(Additional graph paper, if required, can be found on *Page thirty-six.*)

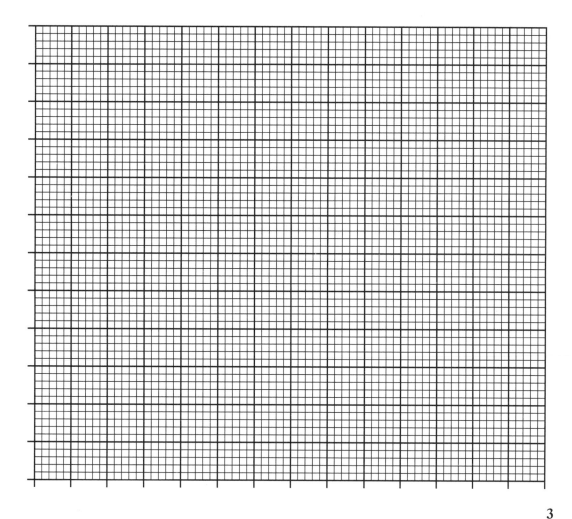

3

(d) What conclusion can be drawn from this investigation?

_____

_____ 1

(e) Suggest a reason why the rate of glucose production is not constant throughout the investigation.

_____

_____ 1

(f) How could the student improve the reliability of his results?

_____

_____ 1

Marks

4. **(continued)**

(g) Some people who have problems digesting lactose are said to be lactose intolerant.

They cannot produce the enzyme lactase.

(i) What general phrase describes an inherited disorder in which the absence of an enzyme prevents a chemical reaction from happening?

_____ 1

(ii) A test can be carried out for lactose intolerance.

The individual being tested does not eat for twelve hours and then drinks a liquid that contains lactose. The individual rests for the next two hours during which their blood glucose level is measured at regular intervals.

What results would be expected if the individual is lactose intolerant?

_____

_____ 1

Marks

5. The diagram below shows a section of a woman's breast shortly after she has given birth.

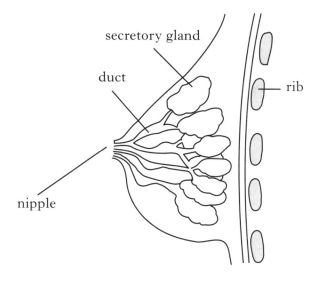

(a) (i) Name the hormone that stimulates the secretory glands to start producing milk.

_____   **1**

(ii) The cells lining the secretory glands are particularly rich in ribosomes. Suggest a reason for this.

_____

_____   **1**

(b) Fluid is not usually released from the breast until the baby suckles.

(i) What name is given to the first fluid that the baby receives from the breast?

_____   **1**

(ii) Describe **one** way in which this first fluid differs from the breast milk produced a few days later.

_____

_____   **1**

(iii) Suckling and crying are examples of non-verbal communication used by a baby. Why is non-verbal communication important to **both** the mother and baby?

_____

_____   **1**

Marks

6. The flow diagram below summarises what happens in the body after a meal of fish and chips.

```
┌─────────────────────────────────────┐
│  Digestion of fish and chips in the  │
│     stomach and small intestine      │
└─────────────────────────────────────┘
                   │
                   ▼
┌─────────────────────────────────────────┐
│ Absorption of the products of digestion  │
│   through the walls of the small         │
│            intestine                     │
└─────────────────────────────────────────┘
                   │
                   ▼
┌─────────────────────────────────────┐
│    Metabolism of some absorbed       │
│      substances by the liver         │
└─────────────────────────────────────┘
                   │
                   ▼
┌─────────────────────────────────────┐
│  Transport of some products of       │
│  metabolism around the body in the   │
│            bloodstream               │
└─────────────────────────────────────┘
```

(a) Explain how bile salts aid the digestion of the fish and chips.

_____

_____ 1

(b) The products of fat digestion are fatty acids and glycerol.

Describe the route taken by these products as they move from the small intestine to the bloodstream.

_____

_____

_____

_____ 2

Marks

6. **(continued)**

(*c*) During the absorption and metabolism of this meal, samples of blood from the hepatic portal vein and the hepatic vein were tested for glucose and urea.

Complete each row of the table below, using the words **Higher** and **Lower**, to compare the concentration of each substance in the two blood vessels.

	Blood vessel	
Substance	Hepatic portal vein	Hepatic vein
Glucose		
Urea		

2

(*d*) State **one** feature of veins which helps to maintain blood flow.

_____

_____   1

(*e*) Drugs and alcohol pass into the bloodstream through the digestive system.

The liver converts these harmful substances into harmless products.

What term describes this action of the liver?

_____   1

[Turn over

Marks

7. A long distance runner took part in some laboratory tests using a treadmill.

She was asked to use the treadmill at a setting of 4 km/h for three minutes during which her pulse rate was monitored. At the end of this time a blood sample was taken which was tested for lactic acid concentration. The procedure was then repeated a number of times at faster speeds.

The results of the tests are shown in the graph below.

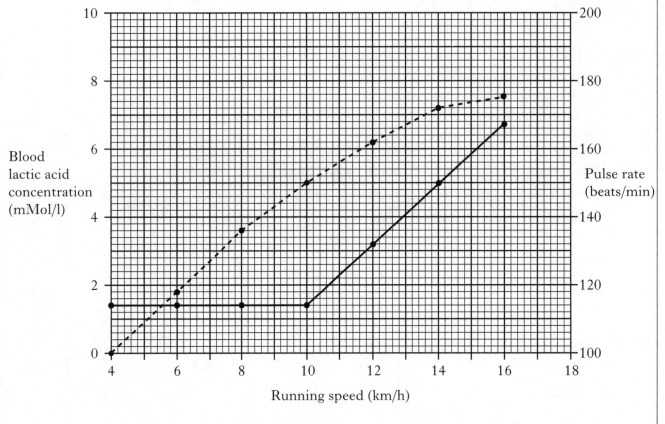

(a) (i) What was the runner's pulse rate when she was running at 6 km/h?

_____ 1

   (ii) State the concentration of lactic acid in the runner's blood when her pulse rate was 172 beats/min.

_____ mMol/l 1

   (iii) Predict what the runner's blood lactic acid concentration would be if she ran at 18 km/h for three minutes.

Blood lactic acid concentration _____ mMol/l 1

DO NOT WRITE IN THIS MARGIN

Marks

7. (continued)

(*b*) A build-up of lactic acid in muscles leads to fatigue.

(i) Explain why lactic acid builds up in the muscles as running speeds increase.

____

____

____

____ **2**

(ii) Distance runners often monitor their pulse rate while they are training.

Suggest how this runner could use a pulse rate monitor and the information from the graph to allow her to run for long periods of time without developing muscle fatigue.

____

____

____

____ **2**

[Turn over

Marks

8. Two men (P and R) were being tested for *diabetes mellitus*, a condition which results in failure to control blood glucose concentration.

    After fasting overnight, they were given a large glucose drink. Their blood glucose concentration was measured immediately (0 hours) and then every hour for five hours.

    The results of the tests are shown in the table below.

	Time after drinking glucose (hours)					
	0	*1*	*2*	*3*	*4*	*5*
Blood glucose concentration of P (mg/100 ml)	145	210	190	180	170	160
Blood glucose concentration of R (mg/100 ml)	90	125	90	85	90	90

(a) It was concluded that P had diabetes and R did not.

   (i) State **two** ways in which the test results indicate that P has diabetes.

   1 _____

   2 _____   1

   (ii) Name the hormone responsible for the change in the blood glucose concentration of R

   (A) between 1 and 2 hours   _____

   (B) between 3 and 4 hours.   _____   1

(b) *Diabetes insipidus* can be caused by a lack of ADH in the body.

   (i) Which organ of the body releases ADH?

   _____   1

   (ii) State an effect that failure to produce ADH would have on the body.

   _____

   _____   1

Marks

9. The diagram below shows a synapse between two nerve cells in the brain and a magnified view of a receptor called NMDA.

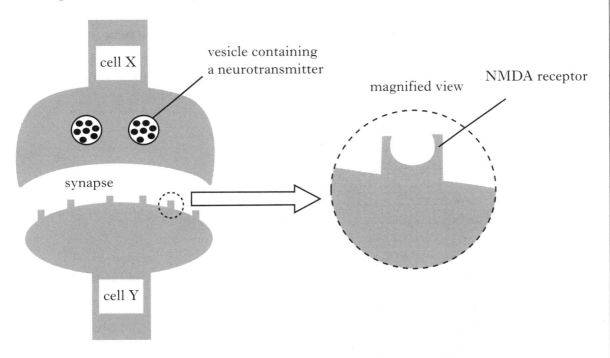

(a) (i)   Describe how the neurotransmitter in the vesicle reaches cell Y.

_____

_____

_____   2

(ii) The diagram above shows a single neural pathway.

Explain how a converging neural pathway would be more likely to generate an impulse in nerve cell Y.

_____

_____

_____   2

(b) Many factors can lead to memory loss.

(i)   One of these factors is a reduction in the number of NMDA receptors.

Which part of the brain contains nerve cells rich in NMDA receptors?

_____   1

(ii) Another factor is the decreased production of acetylcholine.

Name the condition which results from the loss of acetylcholine-producing cells in the brain.

_____   1

DO N
WRIT
THI
MAR(

Marks

10. A study was carried out to compare the influence of genetics with that of the environment on the development of two behavioural conditions, A and B.

   Several hundred pairs of children, from the same families, took part in the study. Some pairs were monozygotic twins, some pairs were dizygotic twins and some pairs were adopted and unrelated.

   In each pair, one of the children had one of the behavioural conditions and investigators observed whether or not the other child shared the condition.

   Results of the study are shown in the bar graph below.

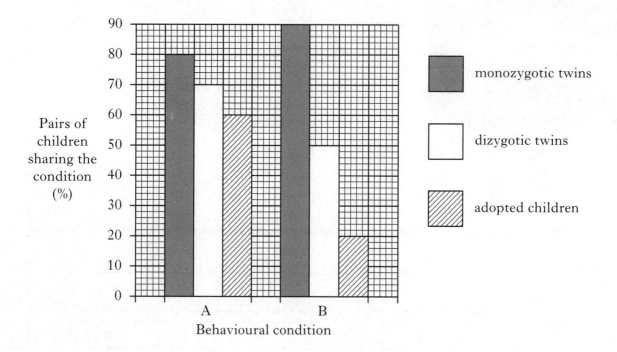

   (a) Explain why it was important that monozygotic twins were chosen for this study.

   _____

   _____

   _____   2

   (b) Use the graph to explain whether conditions A and B are more likely to be caused by genetic or environmental factors.

   (i) Likely cause of condition A _____

   Explanation _____

   _____   1

   (ii) Likely cause of condition B _____

   Explanation _____

   _____   1

Marks

11. The bar graph shows population changes in Scotland for different age groups between 1991 and 2000.

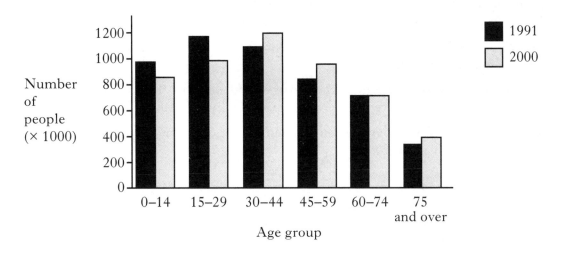

(*a*) Suggest a reason for the population change in those aged 75 and over.

_____

_____ 1

(*b*) Describe **two** ways in which the data for the year 2000 would be different if it were taken from a developing country with a similar population size to Scotland.

1 _____

_____

2 _____

_____ 1

(*c*) Describe **two** ways in which the information in the graph could be used by authorities to plan for the future.

1 _____

_____

2 _____

_____ 1

[Turn over

Marks

12. An investigation was carried out into the influence of adults on the behaviour of young children.

Some groups of children watched a recording of either a man or a woman being physically and verbally aggressive to a large plastic clown.

Other groups of children watched either a man or a woman behaving in a non-aggressive manner towards the clown.

Each child was then placed in a room on their own with the clown. The number of aggressive acts they committed over a five minute period was counted.

The figures in the table below show the average number of aggressive acts that the children committed while in the room.

Gender of children	*Average number of aggressive acts committed by the children*			
	Aggressive man observed	*Aggressive woman observed*	*Non-aggressive man observed*	*Non-aggressive woman observed*
Boys	18·7	7·9	1·0	0·6
Girls	4·4	9·2	0·2	0·8

(a) (i) Which adult/child combination resulted in the least aggression?

_____ 1

(ii) Calculate the percentage increase in aggressive acts committed by boys when they observe an aggressive man rather than a non-aggressive man.

Space for calculation

_____ % 1

(iii) State a conclusion that can be drawn from these results regarding the gender of the aggressive adult.

_____

_____ 1

(b) The children are observing and then repeating the acts of adults. What form of learning are they using?

_____ 1

(c) Suggest a control that could have also been used in this investigation.

_____

_____ 1

13. The graph below shows the application rates of nitrogen and phosphorus to crops in an area of Scotland between 1986 and 2006.

Marks

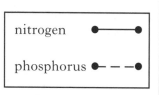

(a) Describe **one** similarity and **one** difference in the data for nitrogen and phosphorus application rate between 1986 and 2006.

Similarity _____

_____

Difference _____

_____   **2**

(b) Express, as a simple whole number ratio, the application rate of nitrogen compared to phosphorus in 1986.

Space for calculation

_____ : _____   **1**
nitrogen   phosphorus

(c) In recent years, there has been a decrease in the use of nitrogen and phosphorus on farms in Scotland.

(i) Suggest **one** way in which this decrease might benefit the environment.

_____

_____   **1**

(ii) Suggest **one** way in which this decrease might disadvantage farmers.

_____

_____   **1**

Marks

14. Glaciers are large masses of ice on mountains and in cold regions of the world. The graph below shows the average change in glacier thickness around the world between 1955 and 2005.

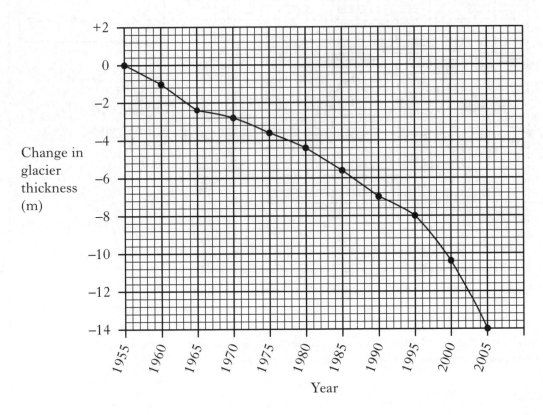

(a) (i) Calculate the average yearly decrease in glacier thickness between 1955 and 2005.

Space for calculation

_____ m/year   **1**

(ii) One consequence of this decrease in glacier thickness is rising sea levels. Describe **one** effect of rising sea levels and subsequent flooding on coastal communities around the world.

_____

_____   **1**

DO NOT WRITE IN THIS MARGIN

Marks

14. (continued)

(*b*) Many people believe that the change in glacier thickness is caused by global warming.

(i) Name **two** gases that contribute to global warming.

1 ______________________ 2 ______________________ **1**

(ii) Give **two** reasons why one of these gases is increasing in the atmosphere.

Gas ______________________

Reason 1 ______________________________________

____

Reason 2 ______________________________________

____ **1**

[Turn over for Section C on *Page thirty-two*

Marks

SECTION C

Both questions in this section should be attempted.

Note that each question contains a choice.

Questions 1 and 2 should be attempted on the blank pages which follow.

Supplementary sheets, if required, may be obtained from the Invigilator.

Labelled diagrams may be used where appropriate.

1. Answer **either** A **or** B.

    A. Discuss memory under the following headings:

        (i) short-term memory;   **5**

        (ii) the transfer of information between short and long-term memory.   **5**

                                (10)

    OR

    B. Discuss how man has attempted to increase food supply under the following headings:

        (i) chemical use;   **4**

        (ii) genetic improvement;   **3**

        (iii) land use.   **3**

                                  (10)

In question 2, ONE mark is available for coherence and ONE mark is available for relevance.

2. Answer **either** A **or** B.

    A. Discuss the biological basis of contraception.   **(10)**

    OR

    B. Discuss the conducting system of the heart and how it is controlled.   **(10)**

[END OF QUESTION PAPER]

[BLANK PAGE]

FOR OFFICIAL USE

Total for
Sections B & C

X009/301

NATIONAL
QUALIFICATIONS
2011

WEDNESDAY, 1 JUNE
1.00 PM – 3.30 PM

HUMAN BIOLOGY
HIGHER

Fill in these boxes and read what is printed below.

Full name of centre

Town

Forename(s)

Surname

Date of birth

Day	Month	Year	Scottish candidate number	Number of seat

SECTION A—Questions 1–30

Instructions for completion of Section A are given on page two.

For this section of the examination you must use an **HB pencil**.

SECTIONS B AND C

1 (a) All questions should be attempted.

    (b) It should be noted that in **Section C** questions 1 and 2 each contain a choice.

2 The questions may be answered in any order but all answers are to be written in the spaces provided in this answer book, **and must be written clearly and legibly in ink**.

3 Additional space for answers will be found at the end of the book. If further space is required, supplementary sheets may be obtained from the Invigilator and should be inserted inside the **front** cover of this book.

4 The numbers of questions must be clearly inserted with any answers written in the additional space.

5 Rough work, if any should be necessary, should be written in this book and then scored through when the fair copy has been written. If further space is required a supplementary sheet for rough work may be obtained from the Invigilator.

6 Before leaving the examination room you must give this book to the Invigilator. If you do not, you may lose all the marks for this paper.

Read carefully

1   Check that the answer sheet provided is for **Human Biology Higher (Section A)**.

2   For this section of the examination you must use an **HB pencil**, and where necessary, an eraser.

3   Check that the answer sheet you have been given has **your name**, **date of birth**, **SCN** (Scottish Candidate Number) and **Centre Name** printed on it.

     Do not change any of these details.

4   If any of this information is wrong, tell the Invigilator immediately.

5   If this information is correct, **print** your name and seat number in the boxes provided.

6   The answer to each question is **either** A, B, C or D. Decide what your answer is, then, using your pencil, put a horizontal line in the space provided (see sample question below).

7   There is **only one correct** answer to each question.

8   Any rough working should be done on the question paper or the rough working sheet, **not** on your answer sheet.

9   At the end of the examination, put the **answer sheet for Section A inside the front cover of this answer book**.

Sample Question

The digestive enzyme pepsin is most active in the

A   stomach

B   mouth

C   duodenum

D   pancreas.

The correct answer is **A**—stomach. The answer **A** has been clearly marked in **pencil** with a horizontal line (see below).

Changing an answer

If you decide to change your answer, carefully erase your first answer and, using your pencil, fill in the answer you want. The answer below has been changed to **D**.

A   B   C   D

SECTION A

All questions in this section should be attempted.

Answers should be given on the separate answer sheet provided.

1. A DNA molecule consists of 4000 nucleotides of which 20% contain the base adenine.

   How many of the nucleotides in this DNA molecule will contain guanine?

   A   800

   B   1000

   C   1200

   D   1600

2. The function of tRNA in cell metabolism is to

   A   transport amino acids to be used in synthesis

   B   carry codons to the ribosomes

   C   synthesise proteins

   D   transcribe the DNA code.

3. The Golgi apparatus is involved in the packaging of

   A   ribosomes

   B   monosaccharides

   C   RNA

   D   enzymes.

4. Which of the following cells secrete antibodies?

   A   B-lymphocytes

   B   T-lymphocytes

   C   Red blood cells

   D   Macrophages

5. The diagram below summarises different types of immunity.

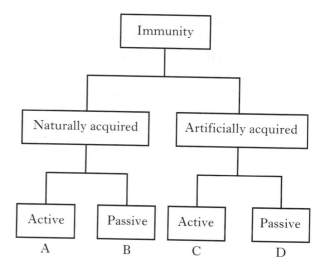

   Which type of immunity would arise from breast feeding?

6. Haemophilia is a sex-linked recessive condition. A woman, who does not have this condition, has a haemophiliac son. The boy's father is also a haemophiliac.

   What are the genotypes of the parents?

	Father	Mother
A	$X^H Y$	$X^H X^h$
B	$X^h Y$	$X^h X^h$
C	$X^h Y$	$X^H X^H$
D	$X^h Y$	$X^H X^h$

[Turn over

7. The table below shows the results of chemical tests on five carbohydrates.

Carbohydrate	Chemical test			
	Iodine solution	Benedict's solution	Barfoed's reagent	Clinistix strip
starch	turns blue-black	stays blue	stays blue	stays pink
sucrose	stays brown	stays blue	stays blue	stays pink
lactose	stays brown	turns orange	stays blue	stays pink
fructose	stays brown	turns orange	turns orange	stays pink
glucose	stays brown	turns orange	turns orange	turns purple

What is the minimum number of tests that would need to be carried out to identify an unknown carbohydrate as lactose?

A   one

B   two

C   three

D   four

8. Huntington's Disease is an inherited condition in humans caused by a dominant allele which is not sex-linked.

A woman's father is heterozygous for the condition and her mother is unaffected.

What is the chance of the woman having the condition?

A   1 in 1

B   1 in 2

C   1 in 3

D   1 in 4

9. The cell shown below is magnified six hundred times. What is the actual size of the cell?

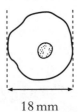

18 mm

A   1080 μm

B   108 μm

C   30 μm

D   3 μm

10. The diagram shows the chromosome complement of cells during the development of abnormal human sperm.

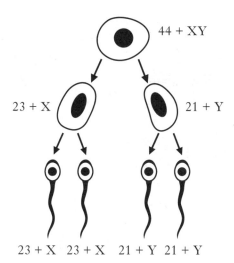

A sperm with chromosome complement 23+X fertilises a normal haploid egg. What is the chromosome number and sex of the resulting zygote?

	Chromosome number	Sex of zygote
A	24	female
B	46	female
C	46	male
D	47	female

11. In fertility clinics, samples of semen are collected for testing.

The table below shows the analysis of semen samples taken from five men.

Semen sample	1	2	3	4	5
Number of sperm in sample (millions/cm³)	40	19	25	45	90
Active sperm (percent)	65	60	75	10	70
Abnormal sperm (percent)	30	20	90	30	10

A man is fertile if his semen contains at least 20 million sperm cells/cm³ and at least 60% of the sperm cells are active and at least 60% of the sperm cells are normal.

The semen samples that were taken from infertile men are

A   samples 3 and 4 only

B   samples 2 and 4 only

C   samples 2, 3 and 4 only

D   samples 1, 2, 4 and 5 only.

12. The graphs below show the hormones involved in the menstrual cycle.

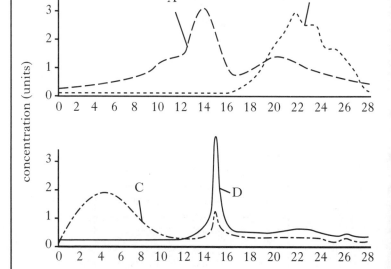

Which line represents oestrogen?

13. The vein in the umbilical cord carries

    A   oxygenated fetal blood

    B   deoxygenated fetal blood

    C   oxygenated maternal blood

    D   deoxygenated maternal blood.

14. A child born to parents with different Rhesus factors can be at risk because

    A   anti-D antibodies from the Rh− mother destroy the baby's red blood cells

    B   anti-D antibodies from the Rh+ mother destroy the baby's red blood cells

    C   anti-D antigens from the Rh− mother destroy the baby's red blood cells

    D   anti-D antigens from the Rh+ mother destroy the baby's red blood cells.

15. The diagram below shows the blood supply to cells lining an air sac in the lungs.

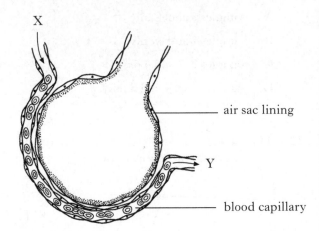

    Which line of the table shows correctly the change in concentration of glucose and oxygen as the blood flows from X to Y?

	Glucose	Oxygen
A	increase	increase
B	increase	decrease
C	decrease	increase
D	decrease	decrease

16. The diagram below shows the liver and its associated blood vessels.

    The arrows show the direction of blood flow.

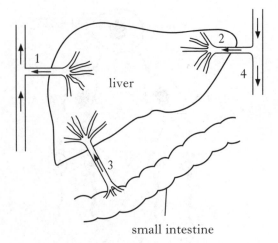

    Which of the blood vessels are veins?

    A   1 and 2

    B   1 and 3

    C   2 and 3

    D   2 and 4

17. Which of the following is **not** a function of the lymphatic system?

    A   Production of tissue fluid

    B   Absorption of products from fat digestion

    C   Removal of bacteria

    D   Production of lymphocytes

18. Which of the following statements about red blood cells is true?

    A   They are manufactured in the liver.

    B   They have a lifespan of 240 days.

    C   Vitamin B_{12} is required for their production.

    D   They are broken down in the kidney.

19. The graph below shows how female bone density changes with age.

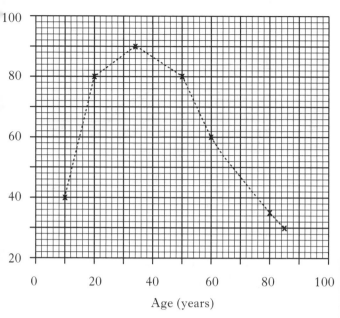

Age (years)

When a female's bone density falls to 60% of its maximum, there is an increased chance of bone breakage.

This occurs at

A   60 years

B   64 years

C   76 years

D   84 years.

20. Mature red blood cells have no nucleus and no mitochondria.

Which of the following processes can be carried out by a mature red blood cell?

A   Aerobic respiration

B   Protein synthesis

C   Anaerobic respiration

D   Cell division

21. Which of the following blood vessels is likely to contain blood with the lowest concentration of urea?

A   Hepatic artery

B   Hepatic vein

C   Renal artery

D   Renal vein

22. Which of the following correctly identifies the locations of the centres that monitor blood water concentration and temperature in humans?

	Blood water concentration	Temperature
A	Hypothalamus	Hypothalamus
B	Hypothalamus	Pituitary gland
C	Pituitary gland	Hypothalamus
D	Pituitary gland	Pituitary gland

23. Infants are more likely to suffer from hypothermia because they have

A   a low surface area to volume ratio

B   a high surface area to volume ratio

C   a low metabolic rate

D   a high metabolic rate.

24. When the body temperature becomes too high, which of the following sets of changes can occur in the skin?

A   Vasoconstriction and contraction of erector muscles

B   Vasodilation and contraction of erector muscles

C   Vasoconstriction and relaxation of erector muscles

D   Vasodilation and relaxation of erector muscles

25. The diagram below shows reactions involved in deamination.

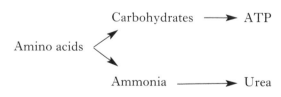

The reactions shown take place in the

A   small intestine

B   kidney

C   gall bladder

D   liver.

26. Which of the following statements is correct?

    A The somatic nervous system controls mainly involuntary actions using sensory nerves.

    B The somatic nervous system controls mainly voluntary actions using sympathetic nerves.

    C The autonomic nervous system controls some involuntary actions using parasympathetic nerves.

    D The autonomic nervous system controls some voluntary actions using motor nerves.

27. A young person does not smoke because she has seen an advertising campaign showing pictures of famous sports stars who do not smoke.

    This is an example of a behaviour called

    A identification

    B discrimination

    C generalisation

    D deindividuation.

28. Which of the following best describes shaping behaviour?

    The reward of behaviour which

    A improves performance in competitive situations

    B approximates to the desired behaviour

    C results in the learning of motor skills

    D results in deindividuation taking place.

29. The graph below shows the birth rate and death rate of a population.

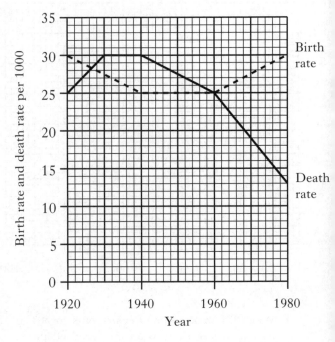

The year of greatest increase in population was

    A 1920

    B 1940

    C 1960

    D 1980.

30. When fertilisers enter a loch, the population of bacteria often increases dramatically.

    Which line in the table below describes correctly the cause of the increase in the bacterial population and the result of the increase in the bacterial population?

	Cause of the increase in population of bacteria	Result of the increase in population of bacteria
A	death of plants	increase in nitrates
B	decrease in oxygen levels	increase in organic matter
C	increase in nitrates	algal blooms
D	increase in organic matter	decrease in oxygen levels

Candidates are reminded that the answer sheet MUST be returned INSIDE the front cover of this answer booklet.

[Turn over for Section B on *Page ten*

Marks

SECTION B

All questions in this section should be attempted.

All answers must be written clearly and legibly in ink.

1. (a) The diagram shows part of an mRNA molecule being formed on a strand of DNA.

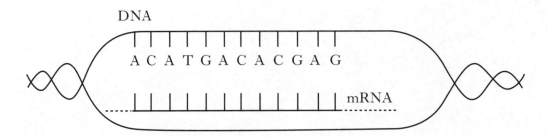

   (i) In which part of the cell is mRNA formed?

   _____   1

   (ii) Complete the mRNA molecule by filling in the correct base sequence **on the diagram.**   1

   (iii) How many amino acids are coded for by this section of mRNA?

   _____   1

   (b) Some diseases are caused when cells in the body produce a harmful protein. Recent research has led to the development of antisense drugs to treat such diseases. These drugs carry a short strand of RNA nucleotides designed to attach to a small part of the mRNA molecule that codes for the harmful protein.

   (i) Suggest how these drugs may prevent the production of a harmful protein.

   _____

   _____   1

   (ii) Antisense drugs can be used to treat autoimmune diseases.

   Describe what is meant by an autoimmune disease.

   _____

   _____   1

DO NOT
WRITE IN
THIS
MARGIN

Marks

2. The diagram below shows a magnified section of the cell membrane of a red blood cell. The numbers show the relative concentrations of potassium ions that are maintained on either side of the membrane.

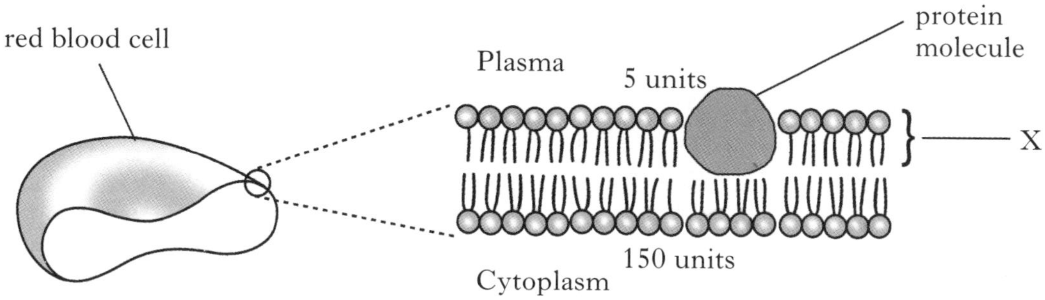

(a) Name molecule X.

_____

1

(b) State **one** possible function of the protein molecule shown in the diagram.

_____

_____

1

(c) Express, as a simple whole number ratio, the concentration of potassium ions inside and outside the cell.

Space for calculation

_____ : _____

inside   outside

1

(d) Use the information in the diagram to name the process by which potassium ions would leave the cell.

_____

1

(e) A sample of blood was treated with a chemical that inhibits respiration.

(i) Describe how this treatment would change the relative concentrations of potassium ions on each side of the membrane.

_____

_____

1

(ii) Explain why the relative concentrations would change.

_____

_____

1

Marks

3. The graph below shows the mass of DNA present as gamete mother cells develop into sperm cells during meoisis in the testes. P and Q represent cells at intermediate stages in this process.

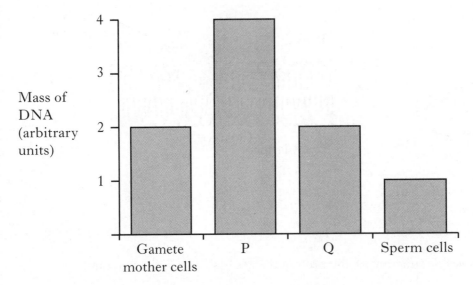

Mass of DNA (arbitrary units)

(a) Explain why the mass of DNA changes between

(i)   the gamete mother cells and cell type P _____

_____ **1**

(ii) cell types P and Q._____

_____ **1**

Marks

3. (continued)

(*b*) The diagram below shows a pair of chromosomes in a cell undergoing meiosis.

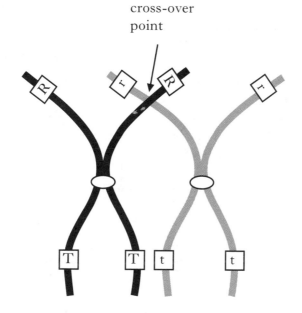

Letters R and T represent dominant alleles of two different genes.

(i) What name is given to the point on the chromosomes where crossing over occurs?

_____   1

(ii) Assuming that crossing over does occur, give all the combinations of alleles that would be present in the resulting gametes.

_____   1

(iii) Crossing over leads to genetic variation.

Name **one** other way in which meiosis increases variation.

_____   1

(*c*) State the exact location of meiosis in the testes.

_____   1

[Turn over

Marks

4. The diagram below shows inheritance of the ABO blood group over three generations of a family. The letters represent the blood group of each individual.

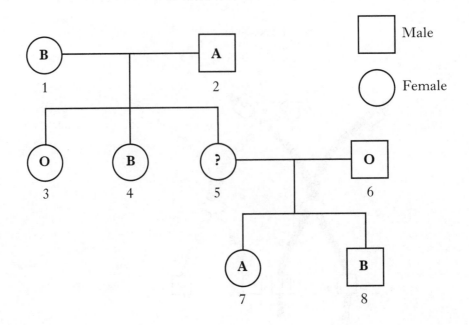

(a) The ABO blood group system is controlled by three alleles: *A*, *B* and *O*. Alleles *A* and *B* are co-dominant and both are completely dominant to allele *O*.

   (i) State the genotypes of the following:

   Individual 1 _____

   Individual 3 _____   1

   (ii) What is the blood group of individual 5? Give a reason for your answer.

   Blood group _____

   Reason_____

   _____

   _____   2

   (iii) How many of the individuals shown in the family tree have a genotype which is heterozygous?

   _____   1

(b) During an operation, individual 4 needed a blood transfusion.

   Identify all the individuals in the family tree who could safely donate blood to her.

   _____   1

Marks

5. The diagrams represent gamete production in an ovary and part of a testis.

Ovary   Testis

(a) (i) Which letter represents a mature ovum?

_____   **1**

(ii) Identify **one** labelled part of each organ which is affected by FSH.

Letter	Name

2

(iii) Describe the effect of testosterone on the testes of an adult.

_____

_____   **1**

(b) Oxytocin is a hormone which is secreted during and after childbirth.

(i) State where oxytocin is produced in the body.

_____   **1**

(ii) Synthetic oxytocin can be used to induce labour.
Describe how it brings about birth.

_____

_____   **1**

[Turn over

DO N
WRITE
THI
MARG

Marks

6. The diagram shows a kidney nephron.

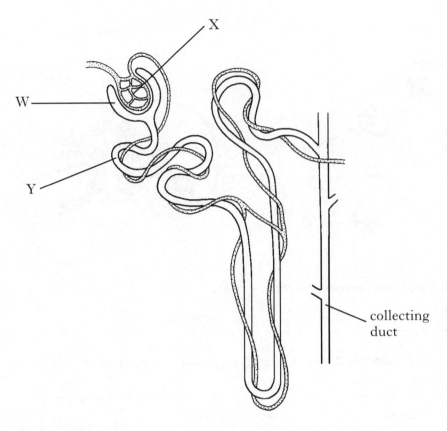

X

W

Y

collecting
duct

(a)   (i)   Name structure W.

_____   **1**

(ii) High blood pressure causes fluid to move from X to W.

Name this process and explain what causes the high blood pressure within X.

Process _____   **1**

Explanation _____

_____   **1**

(b)   (i)   Name structure Y.

_____   **1**

(ii) Describe the main process that occurs in structure Y.

_____

_____   **1**

6. **(continued)**   *Marks*

(c) The graph below shows how changes in the concentration of ADH in the
blood affect the production rate and solute concentration of urine.

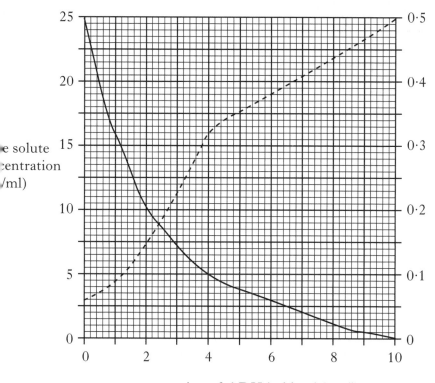

concentration of ADH in blood (mg/l)

(i) **Use the graph** to describe **two** effects of increasing the concentration of
ADH in the blood.

1 _____

_____

2 _____

_____   1

(ii) What is the urine solute concentration when the ADH concentration in
the blood is 6 mg/l?

_____   1

(iii) What is the urine production rate when the urine solute concentration is
4 mg/ml?

_____ litres/hour   1

(iv) If the ADH concentration in the blood remains constant at 4 mg/l,
calculate the mass of solute excreted in the urine in one hour.

Space for calculation

_____ mg   1

Marks

7. (*a*)   The diagram represents a section through the heart.

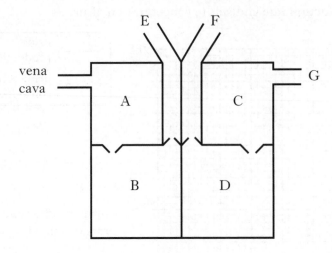

(i)   Name blood vessels E and F.

Vessel E _____

Vessel F _____   1

(ii)   State **two** differences between the composition of the blood in chambers B and D.

1 _____

2 _____   1

(iii)   Place a cross (**X**) on the diagram to indicate the position of the sinoatrial node (SAN).   1

(iv)   Describe the effect of the autonomic nervous system on the sinoatrial node (SAN).

_____

_____

_____

_____   2

(*b*)   State the function of the coronary artery.

_____   1

Marks

8. The table below contains information about diagnosed cases of diabetes in the four countries of the UK in 2008.

Country	Population (million)	Individuals diagnosed with diabetes (% of population)
England	51·3	3·9
Scotland	5·4	3·7
Wales	3·2	4·4
Northern Ireland	1·8	3·4
Total	61·7	

(a) Use the data in the table to calculate the number of individuals in the Scottish population who had diabetes in 2008.

Space for calculation

_____   **1**

(b) A student calculated the percentage of the UK population that had been diagnosed with diabetes by averaging the percentage values in the table. Suggest why this average is likely to misrepresent the true percentage of people in the UK who have been diagnosed with diabetes.

_____

_____   **1**

(c) It has been suggested that the number of people in the UK with diabetes will double by the year 2030.

Suggest **two** different ways in which the current UK government might use this information to plan for the future.

1 _____

_____

2 _____

_____   **1**

[Turn over

DO NO
WRITE
THIS
MARG

Marks

8. **(continued)**

(*d*) The graph below contains information about the number of people in Scotland who consulted their doctor about diabetes in 2008.

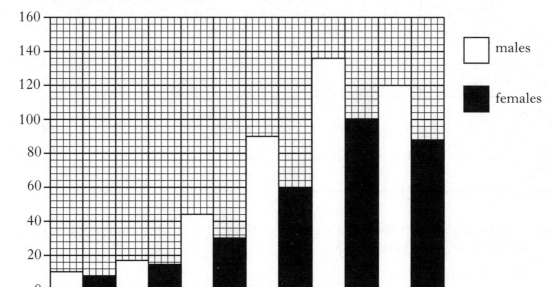

People consulting about diabetes (per 1000 people)

Age group (years)

(i) For people aged between 25 and 74 describe **one** trend shown by the graph which relates to

age _____

_____

gender _____

_____ **1**

(ii) In a Scottish city 2500 men between 45 and 54 years of age visited their doctor in 2008.

Use the graph to calculate how many of these men would be consulting their doctor about diabetes.

Space for calculation

_____ **1**

(iii) Calculate the percentage decrease in the number of men consulting their doctor between the 65-74 age group and the 75+ age group.

Space for calculation

_____ % **1**

Marks

8. **(continued)**

(*e*)   (i)   Type 1 diabetics are unable to produce enough insulin.

Where is insulin produced in the body?

_____

1

(ii)   Describe the role of insulin in the liver.

_____

_____

1

[Turn over

Marks

9. The diagram below represents the passage of information through memory.

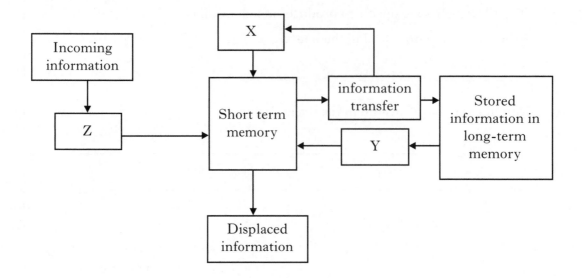

(a) (i) Identify processes X, Y and Z.

X _____

Y _____

Z _____   **2**

(ii) State **two** forms of information which can enter short term memory.

1 _____

2 _____   **1**

(iii) Describe how contextual cues help recall from long-term memory.

_____

_____

_____   **1**

(b) A student had to learn her SQA candidate number which contained 9 digits. She was advised to use chunking to help her memorise it.

Explain why the process of chunking would help her memorise this number.

_____

_____

_____   **1**

DO NOT WRITE IN THIS MARGIN

Marks

9. (continued)

(*c*) (i) Patients with Alzheimer's disease find it difficult to form new memories. Which part of the brain is affected by Alzheimer's disease?

______________________________ **1**

(ii) Name the receptor thought to be important in the process of memory storage.

______________________________ **1**

[Turn over

Marks

10.   An investigation was carried out into the effects of organisation on improving the recall of information.

Four students were each asked to look at a card containing 25 words organised into a branching diagram. The card is shown below.

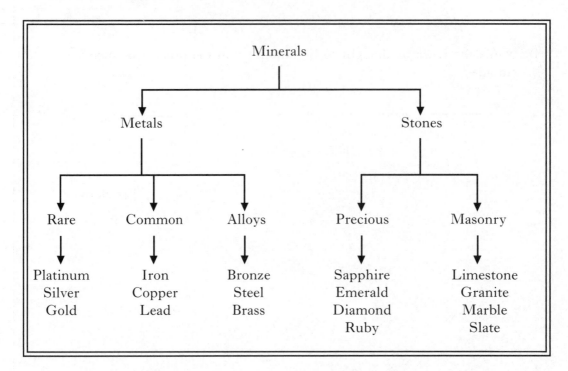

The card was removed after three minutes and each student had to write down as many words as he or she could recall. A score out of 25 was recorded for each student and these were added together to give a total score out of 100 for the group. The procedure was repeated twice. Each time the students were given cards containing 25 different words also organised into branching diagrams. Another group of four students took part in the control for this investigation. The words on their cards were not organised.

The results are shown in the table below.

Student Group	Total number of words recalled (out of 100)			
	1st card	2nd card	3rd card	average
Experimental	75	78	72	
Control	53	57	55	

(a) Complete the table by calculating the average number of words recalled by each student group.

Space for calculation

1

Marks

10. **(continued)**

(b) In what way would the content of the control cards be

similar to the experimental cards? _____

_____

different from the experimental cards? _____

_____   1

(c) Suggest **two** variables, not already mentioned in the description of this investigation, which would have to be kept constant to ensure that a valid comparison could be made between the two groups.

1 _____

2 _____   2

(d) State a conclusion that can be drawn from the results.

_____

_____   1

(e) How could the reliability of the results of this investigation be improved?

_____

_____   1

(f) At the start of the investigation the students were told that the person in each group who recalled most words would be given a prize.

Why did the design of this investigation include a prize?

_____

_____   1

(g) In a further investigation into recall, students were given the same card to memorise on three successive occasions.

Predict what would happen to the number of words recalled on each successive attempt. Explain your prediction.

Prediction _____

Explanation _____   1

[Turn over

Marks

11. The diagram below shows a motor neurone and its junction with skeletal muscle tissue.

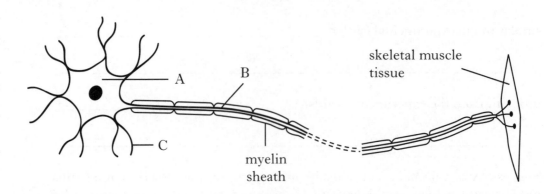

(a) Name the parts of the neurone labelled A, B and C on the diagram.

A _____

B _____

C _____   **2**

(b) Neurotransmitters bind to receptors on skeletal muscle tissue triggering contraction.

   (i) Name **two** neurotransmitters.

   1 _____

   2 _____   **1**

   (ii) Explain why the release of neurotransmitter into a synaptic cleft may sometimes fail to trigger muscle contraction.

   _____

   _____   **1**

   (iii) Name the structural proteins in skeletal muscle tissue and describe how they interact to bring about muscle contraction.

   Proteins _____   **1**

   Description _____

   _____   **1**

DO NOT WRITE IN THIS MARGIN

Marks

11. (continued)

(*c*) (i) State the importance of the myelin sheath in the transmission of impulses.

____

____ **1**

(ii) Post-natal myelination is necessary for a child to go through the sequence of developmental stages leading to walking.

What term describes this sequence of developmental stages?

____________________ **1**

[Turn over

Marks

12. The table below shows the biomass of cod and herring stocks in the North Sea between 1967 and 2004.

The biomass figures are estimates of the total mass of each species present in the North Sea during that year. The critical biomass indicates the mass of each species that must be maintained to prevent it becoming endangered.

Fish species	Estimated Biomass per year (thousand tonnes)					Critical biomass (thousand tonnes)
	1967	1980	1990	2000	2004	
Cod	235	170	75	50	45	150
Herring	920	130	1170	825	1890	1300

(a) (i) Construct a line graph to illustrate the data for cod.

(Additional graph paper, if required, can be found on *Page thirty-four*)

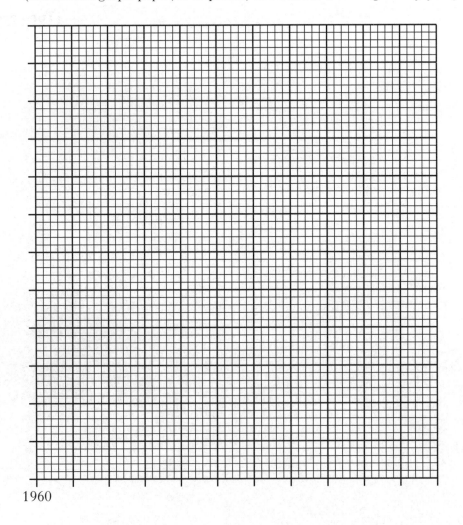

1960

2

(ii) For how many years between 1967 and 2004 was the cod endangered?

Space for calculation

_____ years   1

DO NOT WRITE IN THIS MARGIN

Marks

12. (continued)

(*b*) (i) Calculate the percentage increase in the estimated biomass of herring between 1980 and 1990.

Space for calculation

__________ **1**

(ii) Suggest a reason for the increase in herring biomass between 1980 and 1990.

____

____ **1**

(*c*) What term describes the maximum size of a population which can be sustained by a particular environment?

______________________________ **1**

[Turn over for Section C on *Page thirty*

DO N
WRIT
THI
MAR(

Marks

SECTION C

Both questions in this section should be attempted.

Note that each question contains a choice.

Questions 1 and 2 should be attempted on the blank pages which follow.

Supplementary sheets, if required, may be obtained from the invigilator.

Labelled diagrams may be used where appropriate.

1. Answer **either** A **or** B.

    A. Give an account of communication under the following headings:

        (i) the use of language;   **4**

        (ii) non-verbal communication.   **6**

                                      (10)

    OR

    B. Give an account of the environmental effects of an increasing human population under the following headings:

        (i) deforestation;   **6**

        (ii) increasing atmospheric methane levels.   **4**

                                        (10)

In question 2, ONE mark is available for coherence and ONE mark is available for relevance.

2. Answer **either** A **or** B.

    A. Discuss factors that affect enzyme activity.   **(10)**

    OR

    B. Discuss the production and use of ATP in the body.   **(10)**

[END OF QUESTION PAPER]

SPACE FOR ANSWERS

DO N
WRIT
TH
MAR

SPACE FOR ANSWERS

ADDITIONAL GRAPH FOR QUESTION 12(*a*)(i)

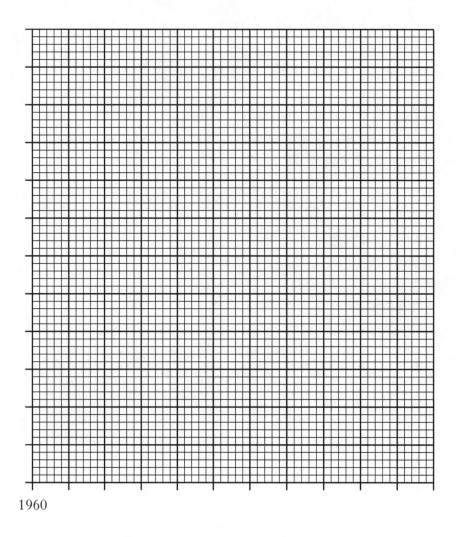

1960

[BLANK PAGE]

FOR OFFICIAL USE

Total for
Sections B & C

X009/12/02

NATIONAL
QUALIFICATIONS
2012

WEDNESDAY, 23 MAY
1.00 PM – 3.30 PM

HUMAN BIOLOGY
HIGHER

Fill in these boxes and read what is printed below.

Full name of centre

Town

Forename(s)

Surname

Date of birth

| Day | Month | Year | Scottish candidate number | Number of seat |

SECTION A—Questions 1–30

Instructions for completion of Section A are given on page two.

For this section of the examination you must use an **HB pencil**.

SECTIONS B AND C

1 (a) All questions should be attempted.

   (b) It should be noted that in **Section C** questions 1 and 2 each contain a choice.

2 The questions may be answered in any order but all answers are to be written in the spaces provided in this answer book, **and must be written clearly and legibly in ink**.

3 Additional space for answers will be found at the end of the book. If further space is required, supplementary sheets may be obtained from the Invigilator and should be inserted inside the **front** cover of this book.

4 The numbers of questions must be clearly inserted with any answers written in the additional space.

5 Rough work, if any should be necessary, should be written in this book and then scored through when the fair copy has been written. If further space is required a supplementary sheet for rough work may be obtained from the Invigilator.

6 Before leaving the examination room you must give this book to the Invigilator. If you do not, you may lose all the marks for this paper.

Read carefully

1   Check that the answer sheet provided is for **Human Biology Higher (Section A)**.

2   For this section of the examination you must use an **HB pencil**, and where necessary, an eraser.

3   Check that the answer sheet you have been given has **your name**, **date of birth**, **SCN** (Scottish Candidate Number) and **Centre Name** printed on it.

     Do not change any of these details.

4   If any of this information is wrong, tell the Invigilator immediately.

5   If this information is correct, **print** your name and seat number in the boxes provided.

6   The answer to each question is **either** A, B, C or D. Decide what your answer is, then, using your pencil, put a horizontal line in the space provided (see sample question below).

7   There is **only one correct** answer to each question.

8   Any rough working should be done on the question paper or the rough working sheet, not on your answer sheet.

9   At the end of the examination, put the **answer sheet for Section A inside the front cover of this answer book**.

Sample Question

The digestive enzyme pepsin is most active in the

A   stomach

B   mouth

C   duodenum

D   pancreas.

The correct answer is **A**—stomach. The answer **A** has been clearly marked in **pencil** with a horizontal line (see below).

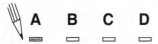

Changing an answer

If you decide to change your answer, carefully erase your first answer and, using your pencil, fill in the answer you want. The answer below has been changed to **D**.

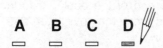

SECTION A

All questions in this section should be attempted.

Answers should be given on the separate answer sheet provided.

1. The diagram below shows some protein filaments in muscle. Which protein is labelled with the letter P?

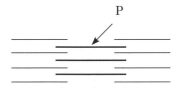

   A   Actin

   B   Adenine

   C   Adenosine

   D   Myosin

2. The following cell components are involved in the synthesis and secretion of an enzyme.

   1   Golgi apparatus

   2   Ribosome

   3   Cytoplasm

   4   Endoplasmic reticulum

Which of the following identifies correctly the route an amino acid molecule would follow as an enzyme is synthesised and secreted?

   A   3 2 1 4

   B   2 4 3 1

   C   3 2 4 1

   D   3 4 2 1

3. How many adenine molecules are present in a DNA molecule of 4000 bases, if 20% of the base molecules are cytosine?

   A   400

   B   600

   C   800

   D   1200

4. The following statements refer to respiration:

   1   Carbon dioxide is released

   2   Occurs during aerobic respiration

   3   The end product is pyruvic acid

   4   The end product is lactic acid

Which statements refer to glycolysis?

   A   1 and 4

   B   2 and 3

   C   1 and 3

   D   2 and 4

5. The diagram below represents a cross-section of a membrane magnified 2 million times.

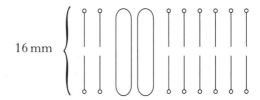

What is the actual width of the membrane?

$1 \, nm = 1 \times 10^{-6} \, mm$

   A $1 \cdot 6 \, nm$

   B $3 \cdot 2 \, nm$

   C $8 \cdot 0 \, nm$

   D $16 \cdot 0 \, nm$

6. During the manufacture of protein in a cell, the synthesis of mRNA occurs in the

   A   nucleus

   B   ribosomes

   C   Golgi body

   D   endoplasmic reticulum.

[Turn over

7. The following diagram shows some stages in the synthesis of part of a polypeptide.

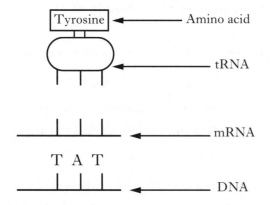

Identify the triplet codes for the amino acid tyrosine.

	On mRNA	On tRNA
A	ATA	UAU
B	UAU	AUA
C	AUA	UAU
D	ATA	TAT

8. Visking tubing is selectively permeable. In the experiment shown below, to demonstrate osmosis, the following results were obtained.

Initial mass of visking tubing
+ contents   = 10·0 g

Mass of visking tubing
+ contents after experiment   = 8·2 g

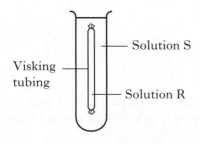

The results shown would be obtained when

A   R is a 5% salt solution and S is a 10% salt solution

B   R is a 10% salt solution and S is a 5% salt solution

C   R is a 10% salt solution and S is water

D   R is a 5% salt solution and S is water.

9. In the formation of gametes when does DNA replication occur?

A   At the separation of chromatids

B   As homologous chromosomes pair

C   Before the start of meiosis

D   At the end of the first meiotic division

10. Identical twins can result from

A   a haploid egg fertilised by a single sperm

B   a haploid egg fertilised by two identical sperm

C   a diploid egg fertilised by a single sperm

D   two haploid eggs fertilised by two identical sperm.

11. The diagram below represents a stage in the division of a cell.

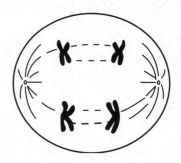

Which line of the table identifies correctly the type of division and the number of chromosomes?

	Type of division	Number of chromosomes
A	mitosis	8
B	mitosis	4
C	meiosis	8
D	meiosis	4

12. Phenylketonuria is caused by a single autosomal gene.

A man and a woman, who are unaffected, have an affected child.

What is the probability that their next child will be affected?

A   25%

B   50%

C   75%

D   100%

13. The offspring from a mother who is homozygous for blood group A and a father who is heterozygous for blood group B, will have a blood group which is

A   AB or A

B   AB or B

C   A or B

D   A or O.

14. A function of the interstitial cells in the testes is to produce

A   sperm

B   testosterone

C   seminal fluid

D   follicle stimulating hormone (FSH).

15. Which of the following is the sequence of events following fertilisation?

A   Cleavage ⟶ Differentiation ⟶ Implantation

B   Implantation ⟶ Differentiation ⟶ Cleavage

C   Differentiation ⟶ Implantation ⟶ Cleavage

D   Cleavage ⟶ Implantation ⟶ Differentiation

16. The graph below shows the changes which occur in a body's food stores during four weeks of food deprivation.

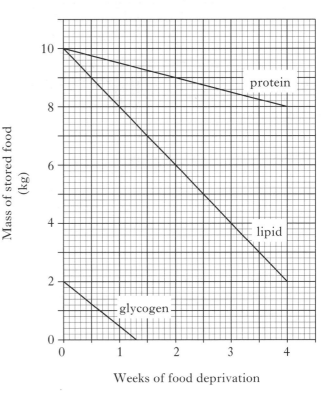

Which of the following conclusions can be drawn from the graph?

A   The glycogen food store decreases at the fastest rate during week one.

B   Between weeks three and four the body gains most energy from protein.

C   Each food store decreases at a constant rate during week one.

D   Between weeks one and four the body only gains energy from lipid and protein.

[Turn over

17. The graph below shows the growth in length of a human fetus before birth.

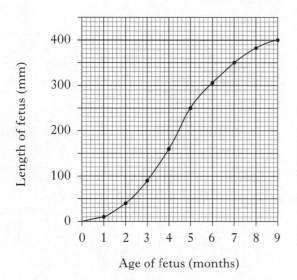

Age of fetus (months)

What is the percentage increase in length of the fetus during the final 4 months of pregnancy?

A   33·3

B   60·0

C   62·5

D   150·0

18. The sperm counts of a sample of men taken between 1940 and 2000 are shown in the graph below.

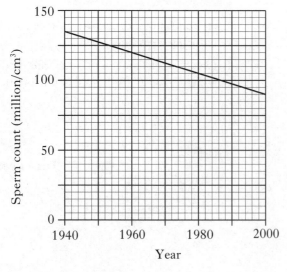

Year

What is the average reduction in sperm count per year?

A   0·67 million/cm^3/year

B   0·75 million/cm^3/year

C   0·92 million/cm^3/year

D   45 million/cm^3/year

19. The effect on the kidney of a high concentration of antidiuretic hormone (ADH) in the blood is to

A   increase tubule permeability which increases water reabsorption

B   decrease tubule permeability which prevents excessive water loss

C   increase glomerular filtration rate which increases urine production

D   decrease glomerular filtration rate which reduces urine production.

20. Compared to the blood in the renal artery, the blood in the renal vein has a higher concentration of

A   oxygen

B   carbon dioxide

C   glucose

D   urea.

21. The graph below records the body temperature of a woman during an investigation in which her arm was immersed in warm water for 5 minutes.

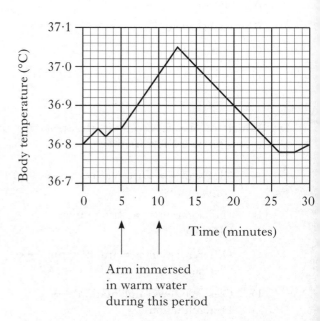

Time (minutes)

Arm immersed in warm water during this period

By how much did the temperature of her body vary during the 30 minutes of the investigation?

A   2·7 °C

B   0·27 °C

C   2·5 °C

D   0·25 °C

22. The flow chart below shows how the concentration of glucose in the blood is regulated.

Blood glucose concentration rises ⟶ Pancreas secretes less of compound X and more of compound Y ⟶ Liver converts glucose to insoluble carbohydrate ⟶ Blood glucose concentration falls

Which line identifies correctly the compounds X and Y?

	Compound X	Compound Y
A	glycogen	insulin
B	insulin	glycogen
C	glucagon	insulin
D	insulin	glucagon

23. The somatic nervous system controls the

   A   skeletal muscles

   B   heart and blood vessels

   C   endocrine glands

   D   muscular wall of the gut.

24. The following is a list of body parts:

   1   tongue

   2   eyebrows

   3   hands

   4   eyes.

Which of these body parts can be used in non-verbal communication?

   A   3 only

   B   2 and 4 only

   C   2, 3 and 4 only

   D   1, 2, 3 and 4

25. An athlete has a much better chance of achieving a "personal best" time in a race rather than in training because of

   A   internalisation

   B   deindividuation

   C   identification

   D   social facilitation.

26. The rewarding of patterns of behaviour which approximate to desired behaviour is called

   A   generalisation

   B   discrimination

   C   extinction

   D   shaping.

27. In the nitrogen cycle, which of the following processes is carried out by nitrifying bacteria?

The conversion of

   A   nitrate to ammonia

   B   nitrogen gas to ammonia

   C   ammonia to nitrate

   D   nitrogen gas to nitrate.

[Turn over

28. The graph below shows the time taken by a student to complete a finger maze, over a number of trials, and the number of errors at each trial.

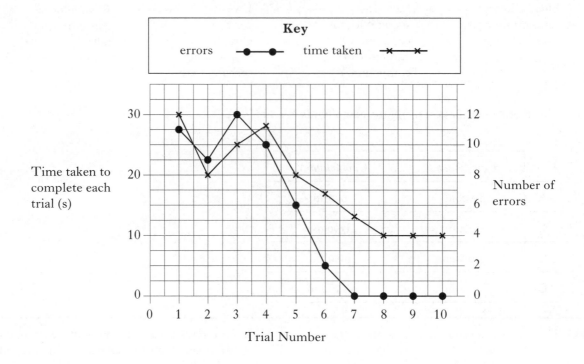

Which of the following statements is correct?

A   The fastest time to complete the maze correctly is 4 seconds.

B   The time taken at trial 5 is 20 seconds.

C   When the number of errors is 10, the time taken is 25 seconds.

D   The number of errors decreased with each subsequent trial.

29. The bar chart below shows the percentage loss in yield of four organically grown crops, as a result of the effects of weeds, disease and insects.

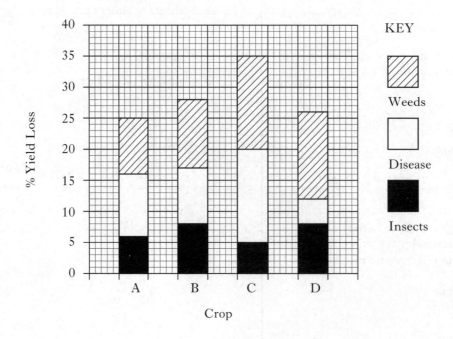

Which crop is likely to show the greatest increase in yield if herbicides and insecticides were applied?

30. The graph below shows how the UK diet changed between 1988 and 1998.

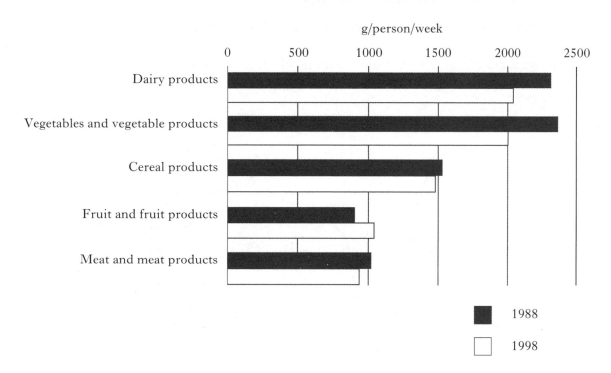

Which of the following conclusions can be drawn from the data?

A   People ate more food in 1998 than in 1988.

B   People ate less food in 1998 than in 1988.

C   People ate a greater variety of food in 1998 than in 1988.

D   People ate a lesser variety of food in 1998 than in 1988.

**Candidates are reminded that the answer sheet MUST be returned
INSIDE the front cover of this answer booklet.**

[Turn over for Section B on *page ten*

SECTION B   *Marks*

All questions in this section should be attempted.

All answers must be written clearly and legibly in ink.

1. The diagram below shows a cell from the lining of a kidney tubule.

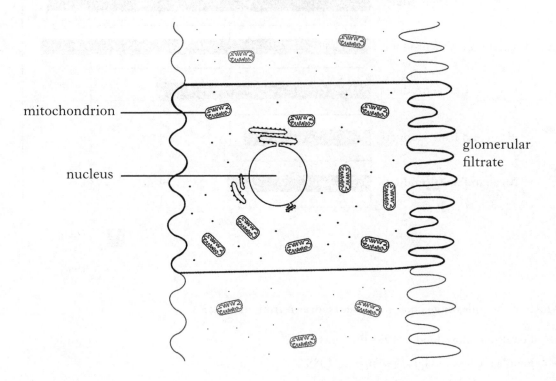

mitochondrion

nucleus

glomerular
filtrate

(*a*) This cell is adapted to reabsorb substances from the glomerular filtrate by
active transport.

(i) What is meant by active transport?

_____

_____ **1**

(ii) Describe how this cell is adapted for active transport.

_____

_____ **1**

(iii) Explain how this cell is adapted for reabsorption.

_____

_____ **1**

Marks

1. **(continued)**

(b) Name the component of the membrane which is involved in active transport.

_____   1

(c) The diagram below shows one of the mitochondria from this kidney tubule cell in greater detail.

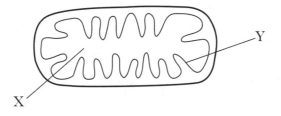

(i) Complete the table below by naming the labelled regions of the mitochondrion and the stage of respiration that occurs there

Region	Name	Respiration stage
X		
Y		

2

(ii) Suggest how the structure of a mitochondrion from a less active cell would differ from the structure of the mitochondrion shown.

Give a reason for your answer.

Structural difference _____

_____

Reason _____

_____   1

[Turn over

Marks

2. The diagram below shows how the immune system responds to a polio virus in a vaccine.

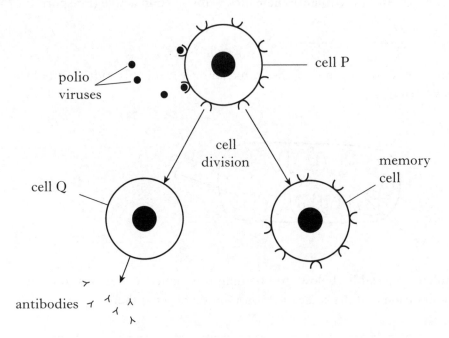

(a) What type of immunological response involves the production of antibodies?

_____   **1**

(b) (i) Name cell Q.

_____   **1**

   (ii) Describe **two** functions of cell P that are shown in the diagram.

   1 _____

   2 _____   **1**

(c) Describe the role of memory cells in the immune system.

_____

_____   **1**

DO NOT WRITE IN THIS MARGIN

Marks

2. (continued)

(*d*) Explain why vaccination against polio would not provide immunity against the measles virus.

____

____ **1**

(*e*) In an emergency, ready-made antibodies can be injected into an individual.

(i) Name the type of immunity that this gives.

______________________ **1**

(ii) State **one** advantage and **one** disadvantage of this type of immunity.

Advantage ______________________________________

____

Disadvantage ___________________________________

____ **2**

[Turn over

Marks

3. Duchenne's muscular dystrophy is an inherited condition in which muscle fibres gradually degenerate.

The condition is sex-linked and caused by a recessive allele.

The family tree below shows the inheritance of the condition through three generations of a family.

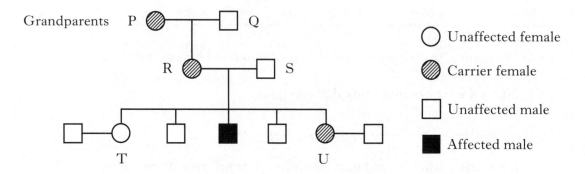

(a) (i) Using the symbols **D** and **d** to represent the alleles, state the genotypes of individuals R and S.

R _____   S _____   1

(ii) What percentage of the grandsons have muscular dystrophy?

_____   1

(iii) Sisters T and U each go on to have a son.

For each sister, state the percentage chance of her son having muscular dystrophy.

Son of T _____   Son of U _____   1

Marks

3. **(continued)**

(b) In humans there is a gene which codes for the essential muscle protein dystrophin.

When this gene is altered, dystrophin is not produced.

An individual with Duchenne's muscular dystrophy cannot make dystrophin.

(i) What general term is used to describe a gene alteration?

_____ 1

(ii) How might the structure of the gene which codes for dystrophin be altered?

_____

_____ 1

(iii) Why does this altered gene fail to produce dystrophin?

_____

_____ 1

(c) Where conditions such as Duchenne's muscular dystrophy exist in a family, the family history can be used to determine the genotypes of its individual members.

What term is used for this process?

_____ 1

[Turn over

Marks

4. (a) Photographic film consists of a clear sheet of plastic coated with chemicals that give it a dark appearance. The chemicals are stuck to the plastic by the protein gelatine.

An investigation was carried out using photographic film and the enzyme trypsin which digests protein.

A piece of photographic film was placed in a test tube containing a solution of trypsin, as shown in **Figure 1** below.

The time taken for the film to turn clear was measured.

The procedure was then repeated using different concentrations of trypsin solution.

The results of the investigation are shown in **Table 1** below.

Figure 1

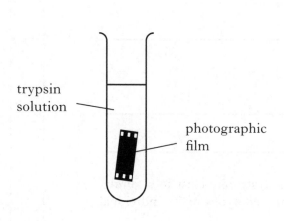

Table 1

Trypsin concentration (%)	Time taken for film to clear (s)
1	112
2	102
3	93
4	84
5	84
6	84

(i) Explain why the photographic film turns clear in this investigation.

_____

_____ 1

(ii) List **two** variables which would have to be kept constant throughout the investigation.

1 _____

2 _____ 2

(iii) How could the reliability of the results of this investigation be improved?

_____

_____ 1

Marks

4. (a) (continued)

(iv) Plot a line graph to illustrate the results of the investigation.

(Additional graph paper, if required, can be found on *Page thirty-six*)

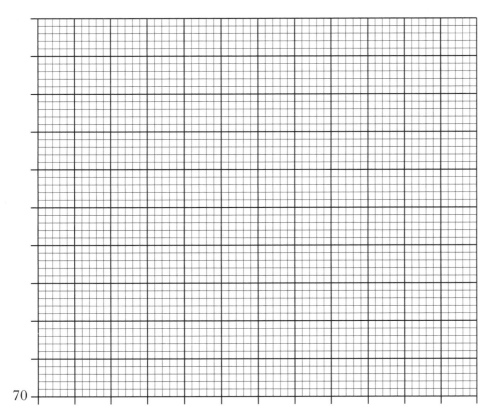

70

2

(v) Explain why the time taken for the film to clear changed as trypsin concentration increased from 1% to 4%.

_____

_____ 1

(vi) Suggest why there was no change in the time taken to clear the film at trypsin concentrations above 4%.

_____

_____ 1

[Turn over

Marks

4. **(continued)**

   (b)   An inactive form of trypsin called trypsinogen is produced and released from the pancreas. Trypsinogen is then converted to trypsin by another enzyme.

       (i)   In which part of the digestive system does activation of trypsin occur?

           _____   **1**

      (ii)   Why are some enzymes such as trypsin produced in an inactive form?

           _____

           _____   **1**

     (iii)   Apart from other enzymes, name another type of molecule that can act as an enzyme activator.

           _____   **1**

Marks

5. The diagram shows a section through the heart and two areas, X and Y, which help to coordinate the heart beat.

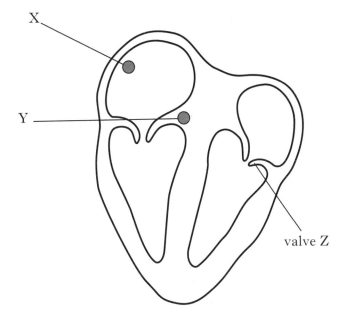

(a) (i) Name structures X and Y.

X _____

Y _____   1

(ii) Electrical impulses travel from X to Y.

What is happening to the heart during this time?

_____   1

(iii) **Draw** arrows on the diagram to show the pathway taken by electrical impulses produced by structure Y.   1

(b) (i) Name valve Z.

_____   1

(ii) During which stage of the cardiac cycle is valve Z closed?

_____   1

[Turn over

Marks

6. The graph below shows the concentrations of two ovarian hormones in a woman's blood during her menstrual cycle.

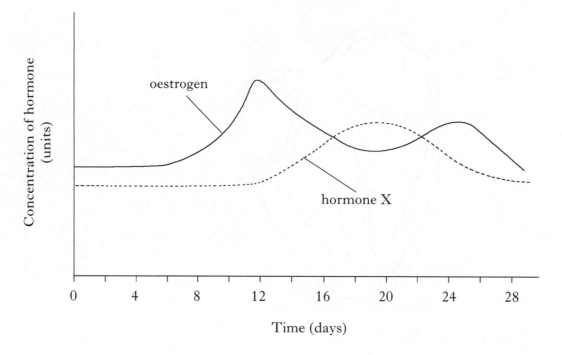

(a) Name hormone X.

_____   1

(b) What effect does oestrogen have on the following structures?

   (i) The uterus between days 4 and 12 in the cycle.

   _____

   _____   1

   (ii) The pituitary gland on day 12 of the cycle.

   _____

   _____   1

(c) Describe **one** way in which the graph would be different if the woman became pregnant during this cycle.

   _____

   _____   1

Marks

6. **(continued)**

(d) The diagrams below show sections through two structures found in the ovary at different times in the menstrual cycle.

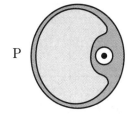 

    (i) Name structures P and Q.

       P _____   Q _____   1

    (ii) What key event in the menstrual cycle occurs before P develops into Q?

      _____   1

[Turn over

Marks

7. The graph below shows changes that occurred in a man's breathing when he inhaled air containing different concentrations of carbon dioxide.

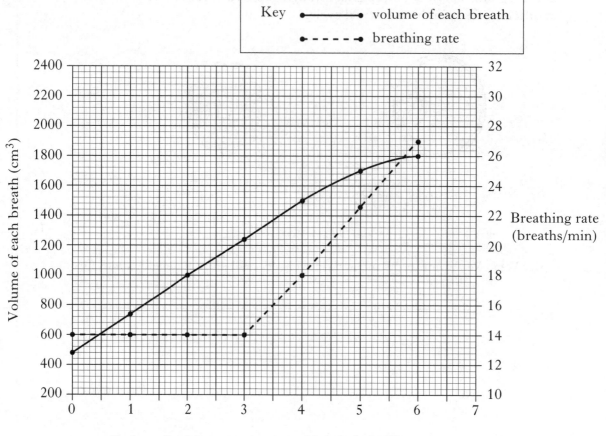

Key •——————• volume of each breath

•– – – – • breathing rate

Carbon dioxide concentration of inhaled air (%)

(*a*) Use data from the graph to describe the changes that occurred in the man's breathing when the carbon dioxide concentration of inhaled air increased from 0 to 3%.

_____

_____

_____

_____

_____   2

(*b*) What was the man's breathing rate when the volume of each breath was 1500 cm^3?

_____ breaths/min   1

DO NOT WRITE IN THIS MARGIN

Marks

7. (continued)

(*c*) Calculate the volume of air inhaled in one minute when the carbon dioxide concentration was 2%.

Space for calculation

____________ cm^3 **1**

(*d*) (i) Predict what the volume of each breath would have been if a carbon dioxide concentration of 7% had been used.

Volume of each breath ____________ **1**

(ii) Suggest why the increase in the volume of each breath becomes less at carbon dioxide concentrations above 4%.

____

____________ **1**

(*e*) On average there is 0·04% carbon dioxide in inhaled air and 4% carbon dioxide in exhaled air.

Explain why this change in carbon dioxide concentration occurs.

____

____________ **1**

[Turn over

Marks

8. The diagram below represents the liver and some associated structures.

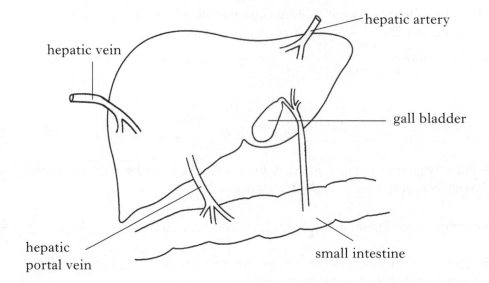

hepatic vein

hepatic artery

gall bladder

hepatic
portal vein

small intestine

(a) **Draw** arrows beside each of the **three** blood vessels to show the direction of blood flow.

1

(b) (i) Name the liquid stored in the gall bladder.

_____

1

(ii) State **one** function of this liquid and explain how it aids digestion.

Function _____

_____

1

Explanation _____

_____

1

(c) Name **one** substance that is stored in the liver.

_____

1

Marks

9. The image below shows a vertical section through a human brain.

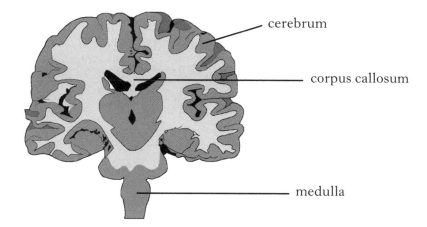

cerebrum

corpus callosum

medulla

(a) Explain how the maximum number of interconnections between neurones is achieved within the cerebrum.

_____

_____

_____   2

(b) What is the function of the corpus callosum?

_____

_____   1

(c) (i) Which division of the nervous system is linked to the medulla?

_____   1

(ii) Describe how this division of the nervous system controls heart rate.

_____

_____   1

[Turn over

Marks

10. The information in the table below refers to the development of walking by infant boys.

Stage of development	Description of behaviour	Age (weeks) at which behaviour develops	
		Earliest	Latest
1	Rolls over	9	23
2	Sits up without support	16·5	32·5
3	Crawls	21	38
4	Pulls up and stands holding on to furniture	23	43
5	Walks holding on to furniture	28·5	49
6	Stands unsupported	35·5	54
7	Walks alone	44·5	57·5

(a) Assuming a normal pattern of distribution, predict by what age 50% of boys would be expected to walk alone.

Space for calculation

_____ 1

(b) Identify all the stages in the development of walking that boys could be at when they are 36 weeks old.

Tick the correct boxes

| 1 ☐ | 2 ☐ | 3 ☐ | 4 ☐ | 5 ☐ | 6 ☐ | 7 ☐ |

1

(c) Suggest **two** reasons why a boy might still only be crawling when, at the same age, his elder brother could stand unsupported.

1 _____

2 _____   1

DO NOT WRITE IN THIS MARGIN

Marks

10. (continued)

(*d*) (i) What term describes the development of a behaviour which follows a set sequence of stages?

________________________________ **1**

(ii) Describe the change which occurs in the nervous system that allows children to go through the stages of development leading to walking.

____ **1**

[Turn over

Marks

11. An investigation was carried out into the effect that the meaning of words has on the ability to recall them from short and long-term memory.

Two groups of people were each shown lists of five words for 30 seconds.

Group 1 was shown words with related meanings while group 2 was shown words with unrelated meanings.

List of words with related meanings – *large, big, great, huge, wide.*
List of words with unrelated meanings – *late, cheap, rare, bright, rough.*

Immediately after the 30 seconds, the people in both groups were asked to write down, in the correct order, the words that they had been shown.

Everyone was then asked to read a book for one hour and told that they would be asked questions about it afterwards.

Instead, after the hour had passed, everyone was again asked to write down, in the correct order, the words that they had been shown in their original list.

The results of the investigation are shown in the table below.

Group	Meaning of words shown	Correct responses immediately after reading the words (%)	Correct responses after reading the book for one hour (%)
1	related	96	54
2	unrelated	96	78

(a) List **two** ways in which the investigators could minimise variation between the two groups of people.

1 _____

2 _____   1

(b) What aspect of memory explains the high percentage of correct responses immediately after reading the words?

_____

_____   1

(c) Suggest why the groups were asked to read a book during the investigation.

_____

_____   1

DO NOT WRITE IN THIS MARGIN

Marks

11. (continued)

(*d*) State **two** conclusions that can be drawn from the results of this investigation.

1 ______________________________

____

2 ______________________________

______________________________ 2

[Turn over

Marks

12. The diagram below shows the changes that affect the population of a country as it undergoes development.

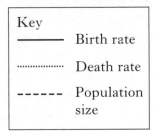

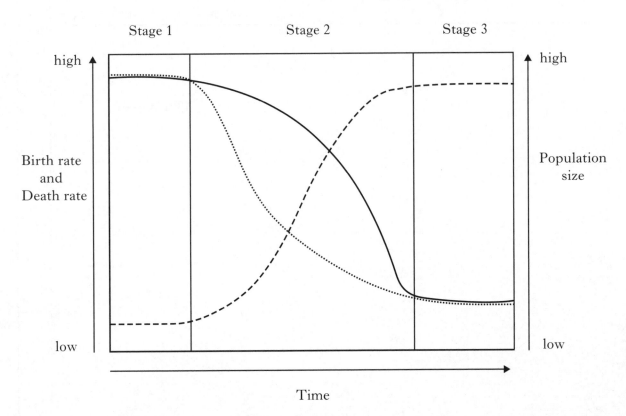

Time

(*a*) (i) Describe the country's birth rate at stage 2 and stage 3 during its development.

_____

_____ 1

(ii) Use information from the diagram to explain why the population size increases rapidly and then starts to level off during stage 2.

_____

_____

_____

_____ 2

Marks

12. (*a*) (continued)

(iii) Suggest **two** factors which may contribute to the change in the death rate during stage 2.

1 _____

2 _____   1

(*b*) The increasing world population requires an increased supply of food.

(i) Pesticides are chemicals which can be used to increase food supply. However, their use can lead to instability in food webs.

Explain this effect.

_____

_____

_____   1

(ii) Other chemicals, such as fertilisers, are also used to increase food production.

Name another method of increasing food production that does not involve chemicals.

_____   1

(*c*) When fertilisers are used in agriculture they can pollute rivers and lochs causing algal blooms.

(i) What is an algal bloom?

_____

_____   1

(ii) Describe the effects an algal bloom might have on a loch.

_____

_____

_____

_____

_____   2

[Turn over for Section C on *Page thirty-two*

SECTION C

Marks

Both questions in this section should be attempted.

Note that each question contains a choice.

Questions 1 and 2 should be attempted on the blank pages which follow.

Supplementary sheets, if required, may be obtained from the Invigilator.

Labelled diagrams may be used where appropriate.

1. Answer **either** A **or** B.

   A   Give an account of the carbon cycle under the following headings:

      (i)   natural uptake and release of carbon;   **4**

     (ii)   disruption of the carbon cycle by human activities.   **6**

                                               (10)

   OR

   B   Give an account of the nervous system under the following headings:

      (i)   the role of neurotransmitters at the synapse;   **6**

     (ii)   converging and diverging neural pathways.   **4**

                                               (10)

In question 2, ONE mark is available for coherence and ONE mark is available for relevance.

2. Answer **either** A **or** B.

   A   Describe the exchange of substances between plasma and body cells.   **(10)**

   OR

   B   Describe involuntary mechanisms of temperature control.   **(10)**

[END OF QUESTION PAPER]

SPACE FOR ANSWERS

SPACE FOR ANSWERS

SPACE FOR ANSWERS

SPACE FOR ANSWERS

ADDITIONAL GRAPH FOR QUESTION 4(*a*) (iv)

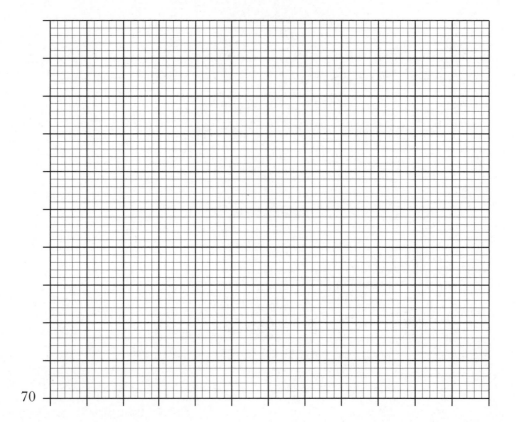

70

HIGHER | ANSWER SECTION

SQA HIGHER
HUMAN BIOLOGY 2008–2012

HIGHER HUMAN BIOLOGY 2008

SECTION A

1. B		16.	C
2. C		17.	A
3. A		18.	C
4. B		19.	A
5. B		20.	D
6. D		21.	C
7. A		22.	A
8. A		23.	A
9. A		24.	D
10. B		25.	B
11. D		26.	B
12. D		27.	C
13. D		28.	C
14. B		29.	C
15. D		30.	D

SECTION B

1. (a) *Any three from:*
   - Thymine (T) in DNA and Uracil (U) in mRNA
   - Deoxyribose in DNA and ribose in mRNA
   - Double strand in DNA and single strand in mRNA
   - DNA found in nucleus, mRNA found in nucleus and cytoplasm
   - DNA can self replicate, mRNA cannot

   (b) 3: Uracil

   8: Guanine

   11: Thymine

   (c) Circle 1, 2, 3 **or** 4, 5, 6 inside or outside nucleus

   (d) Ribosome/rough ER

   (e) X: amino acid/peptide/polypeptide

   Y: tRNA

   (f) (i) Golgi body/apparatus
   (ii) It is packaged/processed/altered/prepared

2. (a) (i)

Protein	Function
B	Transports molecules by diffusion
A	**Acts as an enzyme/catalyst**
D	**Acts as an antigen/cell recognition site/hormone receptor**
C	Transports molecules by active transport

   (ii) X: phospholipid
   Function: Provides flexible/fluid boundary

   (b) Engulfs/invaginates/encircles materials/takes into cell *(direction of movement required)* (one mark)
   Particles/fluid enclosed in vesicle/vacuole/pouch (one mark)

3. (a) AB

   (b) Grandson: BO
   Granddaughter: AO

   (c) One

(d) Each allele has an equal effect/is expressed in the phenotype

(e) True
Group O/son 2 can give to Group A/son 1 *(or vice versa)* **or** O is the universal donor (one mark)
Group A cannot be given to Group O because Group O has anti-a antibodies in the plasma *(or vice versa)* (one mark)

4. (a) pituitary gland

   (b) X: Oestrogen
   Y: Progesterone

   (c) 70

   (d) Day 16 or 17
   Reason: The concentration of LH peaks/high at around this point

   (e) FSH: Stays the same/low **or** decreases/goes down
   Y: Stays the same/high **or** rises/goes up

5. (a) (i) Water: 178·5 dm^3
   Glucose : 175 g
   Urea: 17 g *(units needed)*
   (ii) 99·17 **or** 99·2% **or** 99%
   (iii) The proximal convoluted tubule

   (b) (i) The damaged membrane allows proteins to pass through filter
   or proteins are able to pass from glomerulus into Bowman's capsule
   (ii) Water moves by osmosis from blood to tissue fluid because the blood is more dilute (lack of soluble proteins)

6. (a) P: artery
   Q: arteriole
   R: venule

   (b) The variation is due to the beating of the heart **or** diastolic/systolic pressure

   (c) There are many arterioles/vessels which have smaller diameter, so high friction/high surface area reduces pressure **or** they give more space for blood flow

   (d) Veins have valves to prevent backflow (one mark)
   The movement of the body/muscles squeezes veins (one mark)

7. (a) 80

   (b) He has a low <u>resting</u> heart rate

   (c) *Any three from:*
   At low intensity, fat consumption is higher than carbohydrate consumption **or** At high intensity, fat consumption is lower than carbohydrate consumption (one mark)
   Carbohydrate consumption increases (steeply) for almost the entire range (one mark)
   while fat consumption increases, levels off, then decreases (one mark)
   One correct value for fat + one correct value for carbohydrate (including units for both) (one mark)

   (d) (i) 90–95
   (ii) Because fats have more energy (2x) than carbohydrates **or** because the body's carbohydrate energy store is much less than the fat store **or** because fat stores last longer

   (e) 240

(f) (i) Liver/muscle
   (ii) Glucagon or adrenaline
   (iii) Protein **or** amino acids **or** muscle protein

8. (a) (i) A Central (Nervous System) **or** CNS
     B Peripheral (Nervous System) **or** PNS
     C Autonomic (Nervous System)
     D Spinal Cord

   (ii) *Any one from:*
- Required for conscious control of activities/movement
- Controls skeletal muscle/sensory reception
- Receives/interprets information from body sensors

(b) (i) Corpus callosum
   (ii) To allow space for more nerve cells/nerve cell connections

(c) (i) Medulla (oblongata)
   (ii) They work in opposition/they have opposite effects

   (iii)

Part of the body	Sympathetic effect
Heart	speeds up
Sweat glands	sweat production increases/stimulated
Small intestine	reduces blood flow/activity/peristalsis **or** redirects blood flow away from intestine

9. (a) It flows/leached from farmland/fields/soil into rivers **or** run-off from farmland

(b) (i) 37·5
   (ii) Insecticide is non-biodegradable/cannot be metabolised **or** Insecticide is retained in tissues/not excreted (one mark)
     Each organism eats many more organisms below it in the food chain (one mark)

(c) 100–105 years

(d) *Any two from:*
- Fertiliser/nitrates/manure *Use:* increases yield/ growth in plants
- Herbicides *Use:* kill weeds and reduce competition
- Pesticides *Use:* kill pests which might eat crops
- Fungicides *Use:* kill fungi which might damage crops

(e) Inhibits/denatures

10. (a) 22 = 39%
   and
   28 = 18%

(b)

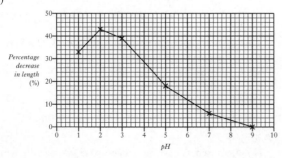

(c) (i) Pepsin works best at a pH 2/optimum pH for pepsin is 2
   (ii) 0

   (iii) Because the starting lengths are different
   (iv) With egg white but no pepsin/pepsin replaced with water
   (v) *Any three from:*
- temperature
- duration
- width of tube
- concentration/volume of enzyme
- type of pepsin
- egg boiled for same time
- type of egg (white)

   (vi) Repeat investigation/test additional pH values

(d) Because it would destroy/digest the cells in which it is produced/destroy the cells lining the stomach

SECTION C

1. A (i) 1. Distinction between voluntary and involuntary *(given in i or ii)*
     2. Curling up to reduce surface area of skin
     3. *Any three from:*
- Increased activity **or** movement
- Additional clothing
- Taking hot drinks
- Switching on heating **or** seeking shelter/warm areas

  (ii) 6. Temperature monitored/responses triggered by hypothalamus
     7. Hair raised to trap air (which is a good insulator)
     8. Vasoconstriction *(must be correctly described in 9 to gain mark here)*
     9. Blood flowing deeper/diverted away from skin surface/vessels smaller to carry heat further away from skin surface
     10. Shivering is contraction of muscles to generate heat
     11. Increase in metabolic rate to generate heat
     12. Decrease in sweat production

  (iii) 13. Hypothermia is a failure of the body temperature regulating mechanisms **or** occurs when body temp falls below a critical level/33, 34, or 35°C
     14. Symptoms of hypothermia
     Any two for one mark from:
- Slowing down of movement
- Difficulty walking – stumbling
- Slurring of speech
- Violent shivering
- Poor vision
- Irrational behaviour/falls asleep/slows down
- False experience of warmth
- Unconsciousness

     15. Reasons for above
     Any two from:
     infants:
- Regulation mechanisms not well developed
- Lack of voluntary responses
- Large surface area: volume ratio

     old:
- Slow metabolic rate
- Inactive
- Regulation mechanisms deteriorating

1. B (i) 1. Growth of facial/pubic/body hair
     2. Growth of bones and muscle accelerated **or** growth spurt takes place
     3. Development of sex organs/testes/penis
     4. Deepening of voice

5. Production of sperm **or** seminal vesicles/prostate activity begins

(ii) 6. LH/ICSH produced by pituitary gland
7. FSH produced by pituitary gland
8. FSH promotes sperm production
9. LH/ICSH promotes testosterone production
10. Testosterone stimulates development of sperm/secondary sexual characteristics
11. Increase in testosterone inhibits pituitary gland **or** ICSH/LH production
12. This is an example of negative feedback control *(must be linked to 11)*
13. Growth hormone from pituitary gland
14. Promotes protein synthesis/elongation of bones/muscle growth
15. Any mention of over/under production leading to dwarfism/gigantism

2. A 1. Controlled by a programme of vaccination
2. Principle of vaccination: weakened pathogen stimulates immune system
3. Control of measles/polio/whooping cough/smallpox *(any two examples)*
4. Control of malaria by use of insecticide/drugs/draining standing water/nets
5. Improved education eg in control of spread of HIV/use of condoms/smoking/breast feeding
6. Improved hygiene/sanitation
7. Provision of clean drinking water eg control of spread of cholera
8. Use of chemicals/sterilising agents eg chlorine in drinking water
9. Effective sewage treatment
10. Improved diet/storing/handling of food eg refrigeration
11. Improved medical facilities/provision of hospitals/doctors *(any two of these)*
12. Work of aid agencies around the world, eg WHO

2. B 1. Maturation
2. An ordered sequence of stages in development: eg standing → crawling → walking
3. Determined by development of nervous system/increased myelination
4. Inherited/genetically determined
5. Any example of inherited condition + brief description of condition (eg Down's)
6. Environmental effects
7. Monozygotic/identical twin studies useful because twins are genetically identical.
8. Influence of parents/family/good/bad upbringing
9. Peer pressure/deindividuation described
10. Social facilitation with description
11. Internalisation **or** identification described
12. Shaping **or** imitation **or** reinforcement/reward/punishment described
13. Generalisation and discrimination described

HIGHER HUMAN BIOLOGY 2009

SECTION A

1.	D	16.	D
2.	B	17.	A
3.	C	18.	B
4.	A	19.	C
5.	C	20.	A
6.	D	21.	A
7.	C	22.	B
8.	A	23.	A
9.	B	24.	B
10.	D	25.	C
11.	A	26.	D
12.	B	27.	D
13.	C	28.	C
14.	C	29.	B
15.	D	30.	B

SECTION B

1. (*a*) (i) Ribose

(ii) Cytosine

(iii) *Any two from:*
Nucleus/cytoplasm/ribosome/<u>rough</u> ER/mitochondria

(*b*) (i) Thymine
Adenine
Cytosine

(ii) ATP/enzymes

(iii) *Any three from four:*
1. <u>mRNA</u> is constructed using DNA as a template/the DNA code/from transcription of DNA
2. <u>mRNA</u> attaches/moves/travels to a ribosome *or* mRNA carries code from nucleus to ribosome/rough ER.
3. <u>Anticodon</u> on <u>tRNA</u> joins with <u>codon</u> on <u>mRNA</u> *or* (Complimentary) <u>base pairing</u> occurs between <u>mRNA</u> and <u>tRNA</u>
4. Order of <u>mRNA</u> codons/bases determines order of amino acids in protein/polypeptide.

2. (*a*) (i) An altered <u>gene</u>/a mutated <u>gene</u>/a faulty <u>gene</u>/<u>a genetic</u> mutation
or
Alteration of base type or sequence or example <u>described</u> (insertion)
or
mutation or alteration of <u>DNA</u>.

(ii) An <u>enzyme</u> is not produced/changed/defective/missing.

(*b*) (i) Enzyme 1 is not produced/does not function *or* Phenylalanine builds up *or* Tyrosine is not produced

(ii) *Any two from:*
Noradrenaline is not produced/production is reduced
Noradrenaline is a <u>neurotransmitter</u>
Phenylalanine builds up <u>and</u> leads to <u>brain</u> damage/impaired <u>brain/mental</u> development

(*c*) Post-natal screening

3. (a) (i)

		mother's gametes	
		M	**N**
father's gametes	**M**	MM	MN
	N	MN	NN

  (ii) 50%

  (iii) Son has only M or N antigens/M or N blood groups
  or
  Son has only one antigen

(b) (i) autoimmunity/autoimmune

  (ii) allergy/allergic reaction

4. (a) Oxygen is produced (becoming trapped in the filter paper causing it to float).

(b) *Any three from:*
  1. size/surface area/diameter/mass/thickness/type of filter paper/disc
  2. concentration of hydrogen peroxide
  3. height/depth/volume of hydrogen peroxide *or* height/depth/volume/size/shape/dimensions of beaker
  4. soaking time
  5. temperature of the solution
  6. pH

(c) Ten discs were used at each concentration *or*
  The experiment was repeated at each concentration **or**
  An average time was taken for each concentration

(d) Use discs soaked in water (added to hydrogen peroxide) **or** use discs containing no catalase (added to hydrogen peroxide)

(e) (i) Correct scales, labels and units on axes
    Points correctly plotted and line drawn

  (ii) 1. As (catalase) concentration increases reaction rate/rate of hydrogen peroxide breakdown increases.
       2. At higher concentrations/above 1% the reaction rate levels off.

(f) Inhibitor binds to the enzyme/catalase altering the shape of the active site
  or
  Inhibitor binds to/blocks the active site of the enzyme/catalase
  or
  Inhibitor competes with the substrate/hydrogen peroxide for the active site

5. (a) X placed or labelled anywhere on the oviduct down to the start of the endometrium.

(b) (i) Corpus luteum.

  (ii) It maintains/thickens/proliferates/prepares the uterus **or** womb lining/endometrium/wall **or**
    It stops menstruation **or**
    It inhibits/prevents ovulation **or**
    It inhibits/reduces FSH/LH production **or**
    It inhibits maturation of the egg/follicle

(c) Oxygen – diffusion
  glucose – active transport
  antibodies – pinocytosis

(d) Cells become specialised/are given specific functions
  or
  Genes are being switched on/off

(e) Cleavage **or**
  Implantation

(f) Identical
  Only one sperm and one egg are involved
  or
  One fertilised egg/zygote (has divided into two)

6. (a) (i) Intrinsic Factor

  (ii) Haemoglobin/cytochrome

(b) 120 days/90-120 days/3 months-4 months/4 months

(c) 27,500 (million)

(d) (i) Biconcave shape provides a large surface area (to volume ratio) **or**
    No nucleus provides more space for haemoglobin

  (ii) Small size/flexible

(e) Spleen/bone marrow

(f) (i) Gall bladder

  (ii) Bile emulsifies lipids/breaks lipids/fat down into smaller globules/pieces/micelles
    This provides a larger surface area for (digestive) enzymes/ lipase
    or
    Bile neutralises stomach acid
    This provides an optimum/suitable pH for (digestive) enzymes/lipase

7. (a) (i) 48 ml/kg/min

  (ii) More/lots of oxygen for respiration/ATP production
    or
    More oxygen/less oxygen debt so lactic acid does not build up

(b) 14·28 litres

(c) Men have larger lungs/a larger heart/larger muscles (than women)

(d) (i) 196

  (ii) 3·6

  (iii) Runner

8. (a) **A** - vena cava
    B – pulmonary artery
    C – carotid artery

(b) The arrows must clearly show the correct direction of movement into and out of the left ventricle.

(c) Prevent backflow of blood into the/left ventricle **or** prevents backflow of blood from the aorta

(d) Ventricular systole

9. (a) (i) In the right atrium.

  (ii) The neurotransmitter binds/attaches/to it **or**
      it acts as a receptor for the neurotransmitter **or**
      To determine whether a signal is excitatory or inhibitory

(b) It is broken down by enzymes **or** enzyme degradation.

(c) (i) The medulla.
  (ii) "Fight or Flight" situation **or**
      Fighting or running (away) **or**
      Any sporting action/exercising **or**
      Any emotional/stress situation (eg excitement/fear/anxiety/anger)

10. (a) (i) **A** – diverging/divergent
    B – converging/convergent

(ii) Impulses go to <u>many/a number of</u> muscles/fingers
This allows fine motor control **or**
This allows coordination of muscles/movements/fingers

(b) (i) An involuntary/automatic/unconscious <u>response/reaction</u>
or
A <u>response/reaction</u> that is not under conscious control/that does not involve thinking

(ii) Plasticity (of response.)

11. (a) 1. As age increases the number of deaths (from lung cancer) increases
2. more males than females die (from lung cancer) after the age of 49/from the age of 50

(b) (i) 45 – 49 years = 1 : 1   60 – 64 years = 5 : 3

(ii) <u>Lung cancer</u> develops faster/earlier in men compared to women.
or
More older men <u>smoke</u> compared to older women <u>while</u> similar numbers of younger men and women smoke.
or
More women stop <u>smoking</u> as they get older (compared to men).
or
More younger women <u>smoke</u> compared to older women.
or
Less younger men <u>smoke</u> compared to older men.

(c) 190

12. (a) (i) A – increase *and*
Sewage provides food for bacteria/contains bacteria
B – decrease *and*
Sewage is diluted/becomes less concentrated/is used up by bacteria **or**
Less food is available for bacteria

(ii) Increase *and*
<u>Decomposition/break down</u> of sewage provides nutrients/nitrates/phosphates **or**
Less cloudy/turbid water so more light/photosynthesis
or
Fertiliser/nitrate/phosphates run-off/leached/washed from farmland

(b) (i) Chemical/substance that kills weeds/plants **or**
A weedkiller.

(ii) Reduces competition for light/water/nutrients/nitrates/phosphates/minerals **or**
Crops get more light/water/nutrients/nitrates/phosphates/minerals

(c) Genetic engineering

SECTION C

1A (i) **The influence of groups**

1. Social facilitation <u>and</u> deindividuation
2. Social facilitation <u>described</u>- performance improves in the presence of others
3. This occurs when competing with others (coactor effect)
or
This occurs when spectators are present (audience effect)
4. Suitable example, showing better performance <u>and</u> the presence of others

5. Deindividuation <u>described</u>- Individual loses personal identity/gains anonymity when in a group
6. Behaviour deteriorates/individual behaves atypically/commits acts they would not do on their own.
7. Suitable example, to include reference to group <u>and</u> poorer behaviour.

(ii) **Influences that change beliefs**

8. Internalisation <u>and</u> identification
9 Internalisation <u>described</u> - individuals change their beliefs/behaviour through persuasion
10 Suitable example described to show change in belief <u>and</u> source of persuasion
11 Identification <u>described</u> - individuals change their beliefs to be like someone they admire
12 Suitable example described to show change in belief/behaviour <u>and</u> focus of admiration

1B (i) **Possible causes of global warming**

1. Carbon dioxide is increasing (in concentration) in the atmosphere
2. This is due to the activities of an <u>increasing</u> human population
3. CO_2/greenhouse gases retain (the sun's) heat causing the greenhouse effect/global warming
4. (Increased) burning of fossil fuels/coal/oil/gas produces CO_2/greenhouse gases
OR
Increased use of fossil fuels by industry/transport releases more CO_2/greenhouse gases
5. Deforestation/removal of trees
6. Means less CO_2 is being absorbed/used by plants for photosynthesis
7. Methane (a greenhouse gas) is released from paddy fields/cattle/landfill sites
8. CFCs (greenhouse gases) are released from aerosols/fridges

(ii) **Potential effects of rising sea levels**

9. (Melting of the polar ice caps/expansion of sea water) causes <u>flooding</u>
10. This will result in homelessness/lack of building land/migration/death of people
11. This will lead to famine/less food production/ fertile land being submerged/destroyed.
12. Habitats for wildlife will be destroyed/example described of a specific animal's habitat being destroyed.
13. Weather patterns will change/description of winds/storms/rainfall etc.

2A 1. <u>Active immunity</u> is when the body makes its own antibodies <u>in response to an infection/pathogen/disease</u>.
2. Invading viruses/bacteria/microbes contain antigens (on their surface).
3. Lymphocytes recognise foreign/non-self antigens (on the invading microbes)
4. B lymphocytes produce antibodies.
5. Antibodies are specific/have receptor sites which bind/attach to foreign antigens.
6. T lymphocytes kill the infected cell/produce chemicals that destroy microbes.
7. Following an infection memory cells are produced/remain in the body.
8. These detect a reinvading microbe and destroy it (before it can cause infection)
9. This (secondary) response is <u>faster/stronger</u> (than the first primary response)
10. <u>Passive immunity</u> occurs when the body gains antibodies (from the mother)

11. A fetus gains antibodies (from the mother) through the
    placenta.
12. Babies gain antibodies from colostrum/breast milk.

2B *Nature*
1. Viruses can only reproduce inside a cell of another
   organism/cannot reproduce without the help of another
   cell/are obligate parasites.
2. Viruses are specific in the type of cell that they attack **or**
   example described.
3. Viruses are transmitted to other people through
   sneezing/coughing/insect bites/sexual
   fluids/contact/drinking.
4. Diseases caused by viruses include the Common Cold,
   Influenza, AIDS, Polio, Smallpox, Hepatitis, Rabies etc
   (must mention 2 diseases)
5. Viruses contain a nucleic acid/DNA/RNA
   surrounded/enclosed by a protein coat/capsid

Reproduction
6. Virus enters host cell **or** injects/inserts DNA/RNA into the
   cell
7. Virus uses cell's ATP/nucleotides/amino acids/enzymes
   (must mention 2 chemicals)
8. (Many) copies of viral nucleic acid/DNA/RNA are
   made/transcribed/replicated
9. (Many) protein coats are formed/translated
10. Nucleic acid/DNA/RNA enters protein coat **or** viruses are
    assembled
11. Host cell bursts/lysis/budding occurs releasing viruses
12. Viruses can then infect other cells/can remain dormant for
    long periods.

HIGHER HUMAN BIOLOGY 2010

SECTION A

1.	C	16.	D
2.	D	17.	B
3.	D	18.	C
4.	B	19.	A
5.	D	20.	B
6.	A	21.	A
7.	C	22.	B
8.	D	23.	C
9.	C	24.	B
10.	D	25.	C
11.	A	26.	A
12.	A	27.	D
13.	C	28.	D
14.	A	29.	B
15.	B	30.	C

SECTION B

1. (a) Meiosis/meiotic (division)

   (b) A = 46   B = 23   C = 23

   (c) B has two chromatids/strands and C has one
       (chromatid/chromosome/strand) **or**
       B is double stranded and C is single stranded

   (d) Independent/random assortment and crossing over

   (e) Seminiferous tubules

2. (a)

Stage	Name	Location
A	Glycolysis	Cytoplasm
B	Krebs/Tricarboxylic/ Citric acid cycle	Matrix of mitochondrion
C	Cytochrome/ Hydrogen/electron transfer system/chain	Cristae of mitochondrion

   (b) Pyruvic acid – 3 (carbons) and citric acid 6 (carbons)

   (c) R is NAD and R transports/delivers/carries hydrogen
       S is Oxygen and S removes hydrogen/acts as the (final)
       hydrogen acceptor/joins with hydrogen to form water

   (d)

Situation	Respiratory substrate	Explanation
Prolonged starvation	Protein/ amino acids	Carbohydrates/ glycogen/glucose and fats/lipids have been used up **or** protein is the only (remaining) energy source/ substrate **or** all other substrates/energy sources used up
Towards the end of a marathon race	Fat/fatty acids/ lipids	Carbohydrate/ glycogen/glucose has been used up

3. (a) (i) P = antigen   Q = antibody
        (ii) B – lymphocyte
        (iii) (A) T-lymphocyte
              makes contact with infected cells <u>and</u> destroys
              them/breaks them down/perforates membrane.
              (B) Macrophage – engulfs/envelops bacteria

   (b) (i) **active** and **naturally**
        (ii) **active** and **artificially**

   (c) Immune system/antibodies attacks body <u>cells</u>
        or Immune system recognises body cells/own antigens as
        foreign/non-self

4. (a) Lactose can be broken down into <u>two</u>
        sugars/monosaccharides/glucose and galactose
        or Lactose is built up from <u>two</u> sugars/
        monosaccharides/glucose and galactose

   (b) *Any two from:*
        1. volume of milk
        2. volume of enzyme/lactase
        3. concentration of enzyme/lactase
        4. temperature of the milk/solution
        5. age of milk

   (c)

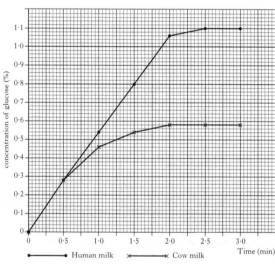

        Correct scales and labels on axes
        Points correctly plotted and lines drawn going to zero
        Lines distinguished from each other (eg key given)

   (d) Human milk contains more <u>lactose</u> (than cow's milk)

   (e) Lactose/substrate is used up/starts to limit the rate of
        reaction

   (f) Repeat experiment <u>and</u> calculate an average

   (g) (i) An <u>inborn</u> error of metabolism/<u>inborn</u> metabolic
             disorder/error
        (ii) (Blood) glucose levels will remain/stay at
             normal/constant concentration/does not rise/stays low

5. (a) (i) Prolactin
        (ii) Synthesis/production of proteins/antibodies

   (b) (i) Colostrum
        (ii) Contains <u>more/many</u> antibodies/protein/ vitamin A **or**
             Contains <u>less</u> fat/lactose/vitamin C **or**
             Colostrum is more yellow/more watery
        (iii) Allows for <u>bonding/attachment</u>

6. (a) Emulsification of <u>fats/lipids/oils</u> **or**
        Breakdown of <u>fats/lipids/oils</u> to smaller droplets/ globules

   (b) Absorbed across/enters <u>villi</u> (of small intestine)
        Passes into <u>lacteal</u>
        Transported through <u>lymphatic system/lymph</u> (to
        bloodstream)

   (c)

Substance	Blood vessel	
	Hepatic portal vein	Hepatic vein
Glucose	**Higher**	**Lower**
Urea	**Lower**	**Higher**

   (d) (The presence of) valves **or**
        Large lumen/diameter/bore (reduces resistance to blood flow)

   (e) Detoxification

7. (a) (i) 118 beats/min
        (ii) 5
        (iii) 8.4/8.5/8.6

   (b) (i) *Any three from:*
             Increased muscle contraction occurs/muscles work
             harder
             This requires more energy/ATP
             Not enough oxygen (reaches muscles to release enough
             energy)/oxygen debt builds up (in muscles)
             (More) <u>anaerobic</u> respiration occurs **or**
             Pyruvic acid not converted to acetyl CoA/pyruvic acid
             converted to lactic acid
        (ii) Use monitor to) keep pulse rate below/at
             <u>150 beats/min</u> when running
             This will keep <u>lactic acid</u> levels low/at 1.4 mMol/l **or**
             This prevents a build up of <u>lactic acid</u>

8. (a) (i) *Any two from:*
             High/higher glucose concentration
             Large/larger increase in glucose concentration
             Glucose concentration decreases slowly/does not
             return to starting value/norm
        (ii) A – Insulin
             B – Glucagon

   (b) (i) Pituitary (Gland)
        (ii) Produce a high volume of urine/increased water loss
             or
             Low concentration of urine **or**
             Dehydration/thirst/low blood pressure/lower water
             concentration in blood **or**
             Less water <u>reabsorbed</u> (in kidney/back into blood) **or**
             No change in permeability of kidney tubules **or**
             Cannot control blood water level/concentration

9. (a) (i) Vesicle fuses/joins with membrane <u>and</u>
             neurotransmitter is released (into synapse/synaptic
             cleft) **or**
             It is released by <u>exocytosis</u> (into synapse/synaptic cleft)
             Neurotransmitter <u>diffuses</u> across the synapse **or**
             Travels across synapse <u>and</u> attaches to receptor
        (ii) Two or more cells/axons/nerve fibres meet one
             cell/cell Y
             More neurotransmitter is released (which
             stimulates/binds to more receptors) **or**
             Threshold is more likely to be reached

   (b) (i) Limbic System/hippocampus
        (ii) Alzheimer's (disease)

10. (a) Monozygotic twins are <u>genetically</u> identical/share same
         <u>genes/DNA</u> **or**
         <u>Genetic</u> factors can be discounted
         Therefore, any difference between them must be due to the
         <u>environment</u>

   (b) (i) Environmental
             Little difference exists between the three groups/pairs
             or
             A high percentage of adopted (unrelated) pairs have
             the condition

(ii) Genetic
The more genetic similarity the greater chance of sharing the condition **or**
A very high percentage of monozygotic twins share the condition <u>and</u> a much lower percentage of other/adopted pairs share it

11. (a) Better healthcare/increased use of vaccination/increased use of antibiotics
Example of medical advance (scanners etc)
Example of a social service (eg meals on wheels, sheltered housing)
Improved diet

(b) 1. More young children/0-14 group would be larger
2. Less old people/older groups would be smaller

(c) *Any two from:*
<u>More</u> health provision/doctors/hospitals (for elderly)
<u>More</u> social provision/residential care/pensions (for elderly)
<u>Less</u> school provision/teachers

12. (a) (i) Non-aggressive man and girls
(ii) 1770
(iii) Children will be more aggressive/likely to copy behaviour if they observe an adult of <u>their own gender/sex</u> (being aggressive)

(b) Imitation

(c) Use children who had not seen the recording/adults with the clown

13. (a) Similarity –
Both (nitrogen and phosphorus application rates) peak in <u>1994</u> **or**
Both increase up to <u>1994</u> <u>and</u> both decrease after (1994) **or** both nitrogen and phosphorus application rates are lower in <u>2006</u> than in <u>1986</u>
Difference –
Nitrogen application rates are <u>always</u> higher than phosphorus application rates **or**
Overall decrease in nitrogen application rate is greater than overall decrease in phosphorus/ nitrogen rate drops faster than phosphorus rate.

(b) 3 : 1

(c) (i) Less algal blooms/less eutrophication/less fertiliser in waterways/less leaching of fertiliser
(ii) Decrease in (crop) yield
Decrease in crop growth/rate of crop growth

14. (a) (i) 0.28
(ii) Erosion/loss of farmland/decreased crop yield/loss of homes/overcrowding/emigration

(b) (i) *Any two from:*
Carbon dioxide/Methane/CFCs/Nitrous oxides/Water vapour
(ii) *Any two from:*
Carbon dioxide – burning fossil fuels/power stations/ transportation/deforestation
Methane – rice fields/cattle/landfill sites/melting permafrost
CFCs – aerosols <u>and</u> fridges/freezers
Nitrous oxides – burning fossil fuels <u>and</u> agricultural soil (nitrification and denitrification)
Water – increased evaporation <u>and</u> plane travel

SECTION C

1A (i) Short term memory
1. Capacity is around 7 pieces of information (+/- 2)
2. This is called the <u>memory span</u>

3. Held for a (very) short period of time/seconds only/30 seconds
4. Chunking increases memory span/capacity/information held
5. Example of chunking <u>described</u> (not just 'eg phone numbers')
6. <u>Serial position</u> effect named <u>and</u> described (or labelled graph)
7. <u>Encoding</u> named <u>and</u> two methods mentioned (acoustic, semantic, visual, smell, taste, touch)

(ii) The transfer of information between short and long-term memory
8. <u>Rehearsal</u> named <u>and</u> described (repetition/ rehearsing of items to be memorised)
9. <u>Organisation</u> named <u>and</u> described (putting items into groups or categories)
10. <u>Elaboration</u> named <u>and</u> described (adding meaning to information)
10a. *mention of all three terms without description*
10b. *mention of all three descriptions without terms*
11. <u>Retrieval</u> named <u>and</u> described (taking information out of long-term memory)
12. <u>Contextual cues</u> aid retrieval/remembering
13. Example of contextual cue given
14. <u>Description</u> of a memory aid (mnemonics/ mind map)

1B (i) Chemical use
1. <u>Fertilisers</u> are used to improve plant growth/provide nutrients for plants
2. <u>Pesticides/insecticides</u> are used to kill/remove pests/insects
3. <u>Herbicides</u> are used to kill/remove weeds
4. Herbicides reduce competition between weeds and crops
5. <u>Fungicides</u> are used to kill fungi/reduce fungal infections
5a. *Three terms (-cides) without descriptions*
6. <u>Antibiotics/growth hormones</u> improve growth of animals

(ii) Genetic improvement
7. Selective breeding (or description)
8. Example of increased yield/increased disease resistance from selective breeding (more grain, more milk etc)
9. Genetic engineering/genetic manipulation/ genetic modification/somatic fusion
10. Definition of genetic engineering as genes being transferred between organisms
11. Result of genetic engineering is increased yield/disease resistance/drought resistance

(iii) Land use
12. Deforestation/description of forest removal
13. Marginal land use described/land reclamation/terracing hillsides
14. Irrigation <u>described</u>
15. Removal of hedgerows/creation of large fields/monoculture use
16. Mechanisation/crop rotation linked to more efficient use of land

2A
1. Contraception is prevention of fertilisation/pregnancy
2. Fertile period lasts for a few days around day 14/mid point of menstrual cycle
3. Fertile period can be detected by <u>rise</u> in body temperature
4. Fertile period can be detected by changes in <u>cervical</u> mucus/mucus becomes thinner

5. Contraceptives can be pills/injections/ implants
6. These contain oestrogen/progesterone
7. Pills usually taken for 3 weeks/one pill taken each day
8. Concentration of hormones (in blood) is increased
9. Causes negative feedback effect/inhibitory effect on pituitary gland
10. Reduced production of FSH prevents maturation of ova/eggs
11. Reduced production of LH prevents ovulation
11a. *mention of reduced production of FSH–LH without functions*
12. (Prolonged/regular) breast feeding/suckling acts as contraceptive

2B
1. Controlled by <u>autonomic nervous system</u>
2. Sympathetic speeds up heart <u>and</u> parasympathetic slows down heart
3. Medulla (oblongata) is control centre (in the brain)
4. Adrenaline speeds up heart rate
5. Pacemaker/SAN <u>in right atrium</u>
6. Pacemaker starts contraction/produces impulses
7. Impulses cause the atria to contract/atrial systole
8. Reaches/stimulates the <u>AVN</u>
9. AVN found at junction of/between atria and ventricles
10. Impulse (from AVN) carried by (conducting) nerves/fibres/bundle of His
11. (Purkinje) fibres/nerves spread out over the ventricles
12. Causes contraction of ventricles/ventricular systole
13. Followed by relaxation/resting/diastolic phase

HIGHER HUMAN BIOLOGY 2011

SECTION A

1.	C	16.	B
2.	A	17.	A
3.	D	18.	C
4.	A	19.	B
5.	B	20.	C
6.	D	21.	D
7.	B	22.	A
8.	B	23.	B
9.	C	24.	D
10.	D	25.	D
11.	C	26.	C
12.	A	27.	A
13.	A	28.	B
14.	A	29.	D
15.	C	30.	D

SECTION B

1. (*a*) (i) Nucleus
　　(ii) U G U　A C U　G U G　C U C
　　(iii) 4

　(*b*) (i) They result in a short/incomplete protein/polypeptide
or
The mRNA cannot bind to the ribosome
or
They prevent translation/tRNA molecules <u>binding</u> to mRNA
　　(ii) The body/immune system/antibodies attacks body <u>cells</u>/own <u>cells</u>
or
A disease in which the body/immune system recognises body cells/self antigens as foreign/non-self

2. (*a*) Phospholipid

　(*b*) Antigen/antigenic marker/enzyme/receptor

　(*c*) 30 : 1

　(*d*) Diffusion

　(*e*) (i) (The concentration) would become equal/closer
or
(The concentration) decreases inside/in the cytoplasm <u>and</u> increases outside/in the plasma
　　(ii) No/less ATP/energy production <u>and</u> no/less active transport
or
Active transport stops/decreases <u>and</u> diffusion equalises the concentrations
or
Respiration is needed to make ATP <u>which</u> is needed for active transport

3. (*a*) (i) DNA/chromosomes <u>replicate</u>
or
Two <u>chromatids</u> are produced (from one chromosome)
　　(ii) <u>Homologous</u> chromosomes/pairs separate
or
Cells go from <u>diploid</u> to <u>haploid</u>

(b) (i) Chiasma/chiasmata
     (ii) RT   Rt   rT   rt
     (iii) Independent assortment

(c) Seminiferous tubules/sperm mother cell

4. (a) (i) 1 – BO
             3 – OO
         (ii) Blood group = AB
              Individual 6 can only pass on allele O to their children and therefore both alleles A and B must have come from person 5 or
              She must provide both A and B alleles to her children as her partner has O alleles/is OO or
              She produced an A and B blood group child with an OO male.
         (iii) 6

   (b) 1, 3, 6 and 8.

5. (a) (i) C
         (ii) A/B – Graafian follicle
                F – Seminiferous tubule/sperm mother cell
         (iii) Stimulates/promotes/causes/increases sperm production

   (b) (i) Pituitary (gland)
         (ii) Stimulates/causes contraction of the uterus/womb

6. (a) (i) Bowman's capsule
         (ii) Process – ultrafiltration
              Explanation – Blood vessel/arteriole entering is wider than blood vessel leaving
                             or
                             Blood flows from an artery/arteriole at high pressure

   (b) (i) Proximal convoluted tubule
         (ii) Reabsorption (or description of reabsorption) of glucose/water/salts/amino acids/vitamins/minerals

   (c) (i) Urine solute concentration increases and urine production rate decreases
         (ii) 19 mg/ml
         (iii) 0.34
         (iv) 1600

7. (a) (i) E – pulmonary artery
             F – aorta
         (ii) Any two from:
              1 – B has more carbon dioxide
              2 – D has more oxygen/D is oxygenated while B is deoxygenated
              3 – B has more glucose
         (iii) X placed anywhere inside the right atrium/A
              or touching outer walls of the right atrium/A
         (iv) Sympathetic increases rate of impulses (from SAN)/heart rate
              Parasympathetic decreases rate of impulses (from SAN)/heart rate

   (b) Supplies the heart (muscle) with oxygen/glucose

8. (a) 199 800

   (b) Not all countries have the same population
        or
        England has a much bigger population
        or
        The population distribution in the (four) countries is not equal

(c) Any two from:
     1. (Increased spending on) provision of doctors/nurses/hospitals/healthcare
     2. (Increased spending on) manufacture of insulin/drugs
     3. (Increased) education/advertising about diet/exercise (and diabetes)
     4. (Increased spending on) research into diabetes
     5. Reduce advertising of sugar-rich foods

(d) (i) Age – More people consult their doctor about diabetes as they get older
          Gender – More males consult their doctor about diabetes than females
     (ii) 110
     (iii) 11.76/11.8/12

(e) (i) Pancreas
     (ii) Insulin causes/stimulates the conversion of glucose into glycogen

9. (a) (i) X – rehearsal
             Y – retrieval
             Z – encoding
         (ii) Visual/Images/sight and Acoustic/Sounds/audio/auditory
              or any reference to two sensory inputs – smell, taste, touch
              or Semantic/Meaning (from LTM) and one sensory input.
         (iii) They serve as a reminder to the time/occasion when the information was originally experienced/encoded

   (b) The nine digits are divided into groups so there are fewer items/chunks to remember.

   (c) (i) Limbic System/hippocampus
         (ii) NMDA

10. (a) 75 and 55

    (b) Similar – cards contain the same words
         Different – words arranged randomly/not organised/not in categories

    (c) Any two from:
         1. Individuals of a similar age range are in each group
         2. Individuals with similar intelligence/ability
         3. Each group has a similar gender/sex balance
         4. Each group has the same time to write down the words
         5. Investigation carried out in same environmental conditions (eg same room/at same time of day/in same temperature/in silence with no distraction)
         6. Words/cards same size/font/colour

    (d) Organisation (of words into branching diagrams) improves their recall/were better remembered

    (e) Repeat the (same) experiment using more/different students/more groups of students

    (f) So all students do their best/are more motivated/concentrate more due to social facilitation/competition.

    (g) Prediction – Number of words recalled increases
         Explanation – Rehearsal/repetition improves transfer into long-term memory (LTM)

11. (a) A – cell body
         B – axon
         C – dendrite

    (b) (i) Acetylcholine and noradrenaline
         (ii) Insufficient secretion/not enough of neurotransmitter
              or
              Threshold level of neurotransmitter not reached

or
Insufficient receptors stimulated

   (iii) Protein – <u>Actin</u> and <u>myosin</u>
      Description – They slide between/over/across each other

(c) (i) Speeds up (the transmission of nerve impulses)
    (ii) Maturation

12. (a) (i)

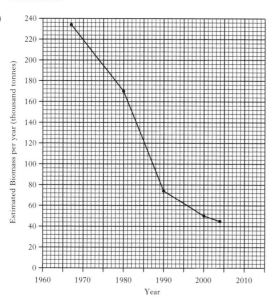

    (ii) 22/23

(b) (i) 800
    (ii) Less disease/decreased predation/increase in food/ fishing quotas or bans introduced/mesh size of fishing nets increased/reduced demand for herring/less competition with cod

(c) Carrying capacity

SECTION C

1A (i) The use of language
1. Language can be written/spoken/uses symbols (to represent information)
2. Language enables information to be organised (into categories/hierarchies)
3. Language allows the transfer of information/skills/ instructions/ideas
4. Language/tone of voice allows communication of views/feelings/moods
5. Language allows learning/intellectual development/cultural development
6. Language allows unique human behaviour/ distinguishes humans from animals
7. Language development is dependent on your environment/imitation/parents

(ii) Non-verbal communication
8. Non-verbal communication is important in early life <u>before speech possible</u>
9. It is important in forming <u>bonds/attachment</u> between infant and parents
10. Example described of non-verbal signal by a baby <u>and</u> the feeling conveyed eg crying for attention
11. Non-verbal communication/body language/facial expression in adults involves signals that they can be <u>unaware</u> that they are giving/<u>subconscious</u> signalling
12. Non-verbal communication/body language/facial expression can aid/replace/ contradict verbal communication **or** sign language use by the deaf **or** signs used for direction/instruction

13. Non-verbal communication/body language/facial expression can signal <u>attitudes/emotions</u>/feelings
14/15. Two examples of non-verbal communication in <u>adults</u> <u>and</u> the feeling/attitude conveyed eg smiling – pleasure, eye contact – attraction, fidgeting – boredom

1B (i) Deforestation
1. Deforestation involves clearing forests for agriculture/building/transport/raw materials/wood/fuel
2. (Cleared) trees/wood burning releases <u>carbon dioxide</u> (into the atmosphere)
3. More carbon dioxide in atmosphere as less <u>photosynthesis</u> occurs
4. <u>Global warming/Greenhouse effect</u> (as less heat escapes from Earth)
5. Loss of <u>roots</u> (which hold/bind soil) causes <u>erosion/ loss of soil</u>
6. (Increased) flooding/silting of rivers/blockage of irrigation systems
7. Deforestation leads to less rainfall/desertification
8. <u>Loss of habitat</u> so species extinction/reduction in numbers occurs **or** reduction in biodiversity

(ii) Increasing atmospheric methane levels
9. (Increased numbers of) cattle/livestock (to feed population)
10. (Increased) growth of rice/paddy fields (to feed population)
11. (Increased) landfill sites (to deal with waste)
12. Methane produced under <u>anaerobic</u> conditions
13. <u>Bacteria</u> produce methane
14. Methane is a greenhouse gas/causes global warming
15. Methane is produced by <u>biomass</u> burning/burning <u>tropical/rain forests</u> (to clear land for farming)

2A
1. Enzymes are catalysts/speed up metabolism/ chemical reaction/lower activation energy
2. <u>Temperature</u>: enzymes have an optimum temperature/temperature at which they work best/work best at 37°C (*or labelled graph to illustrate*)
3. <u>pH</u>: all enzymes have an optimum pH/pH at which they work best (*or labelled graph to illustrate*)
4. <u>Denaturing</u>: a change occurs in the <u>structure/shape/active site</u> of the enzyme at high temperatures/when the pH changes
5. <u>Inhibitors</u>: slow up/stop enzyme activity
6. <u>Competitive inhibitors</u>: attach to/block the active site so keeping out the substrate molecule **or** inhibitor competes with <u>substrate</u> for active site
7. <u>Non-competitive inhibitors</u>: attach to another part of an enzyme and change the shape of the active site/enzyme (so the substrate molecule does not fit)
8. <u>Substrate concentration</u>: <u>increasing</u> substrate concentration <u>increases</u> activity <u>until</u> a point when activity <u>levels off</u> (*or labelled graph to illustrate*)
9. Explanation that activity levels off when all enzyme active sites are reacting with substrate molecules/enzymes are working at fastest rate possible
10. <u>Enzyme concentration</u>: increasing enzyme concentration <u>increases</u> the rate of reaction (*or labelled graph to illustrate*)
11. Explanation that activity increases due to more <u>active sites</u> being added
12. Vitamins/minerals/cofactors/coenzymes/other enzymes <u>activate</u> enzymes

2B

1. ATP is built up from ADP and phosphate (*or equation*)
2. ATP is produced during <u>glycolysis</u>
3. From the conversion of <u>glucose to pyruvic acid</u>
4. ATP is produced from/by the <u>cytochrome system/electron transport chain</u>
5. This is found/takes place on the <u>cristae of the mitochondrion</u>
6. <u>Hydrogen/electrons passed from carrier to carrier</u>, generating (energy to form) ATP
7. Less ATP produced during glycolysis <u>compared</u> to the cytochrome system
8. During anaerobic respiration/lack of/absence of oxygen <u>two</u> molecules of ATP is produced
9. ATP is broken down into ADP and phosphate <u>releasing energy</u> (*or equation*)
10. ATP is produced as fast as it is used up/remains at a constant level in the body
11/12. Uses of ATP – muscle contraction/phagocytosis/protein **or** chemical synthesis/active transport/nerve impulse transmission/glycolysis/sperm swimming/mitosis **or** meiosis **or** cell division/DNA replication

HIGHER HUMAN BIOLOGY 2012

SECTION A

1.	D	16.	C
2.	C	17.	B
3.	D	18.	B
4.	B	19.	A
5.	C	20.	B
6.	A	21.	B
7.	C	22.	C
8.	A	23.	A
9.	C	24.	D
10.	A	25.	D
11.	D	26.	D
12.	A	27.	C
13.	A	28.	B
14.	B	29.	D
15.	D	30.	B

SECTION B

1. (*a*) (i) Movement of molecules/substances/ions against a concentration gradient/from low to high concentration/using energy/using ATP
   (ii) Contains large numbers of /many mitochondria
   or
   Mitochondria provide energy/ATP
   (iii) Folded/convoluted membrane/surface/provides a large/greater/increased <u>surface area</u>

   (*b*) Proteins

   (*c*) (i)

region	name	respiration stage
X	matrix	Krebs/citric/tricarboxylic acid cycle
Y	cristae	Cytochrome system/oxidative phosphorylation/hydrogen or electron transfer system

   (ii) Structure Difference – Mitochondrion would contain fewer folds/cristae
   Reason – Less respiration/ATP/energy is required

2. (*a*) <u>Humoral</u> (response)

   (*b*) (i) B-lymphocyte / plasma cell
   (ii) *Any two from:*
   Attaches/recognises/identifies/detects the (polio) virus
   (Divides to) produce cell Q/ lymphocytes/plasma cells
   (Divides to) produce memory cells

   (*c*) To respond <u>quickly</u> to <u>another/a second</u> invasion of a virus/bacterium/pathogen/toxin/antigen

   (*d*) The measles virus carries different <u>antigens</u> (to the polio virus)
   or
   <u>Antibodies</u> are specific to one virus / polio /antigen
   or
   The <u>receptor</u> on cell P/the B-lymphocyte/the memory cell does not match the measles virus antigen

(e) (i) <u>Artificial passive</u> (immunity)
(ii) Advantage – provides instant/rapid immunity/ protection
Disadvantage – immunity/protection does not last for a long time/ is short-lived/is temporary
or
Memory cells/antibodies are not produced (by body)

3. (a) (i) R = $X^D X^d$ **and**   S = $X^D Y$
(ii) 33 / 33·3 / 33⅓
(iii) Son of T = 0   **and**   Son of U = 50

(b) (i) Mutation
(ii) Alter/change the sequence/order of <u>bases/nucleotides</u>
or
A specific <u>base/nucleotide</u> change is <u>described</u> (insertion, deletion, inversion, substitution <u>described</u>)
(iii) The protein produced contains an altered <u>sequence/ order</u> of <u>amino acids</u>
or
The protein produced contains a <u>different amino acid</u>/is <u>missing</u> an <u>amino acid</u>/has an <u>extra amino acid</u>

(c) <u>Genetic</u> screening/<u>genetic</u> counselling

4. (a) (i) Trypsin/the enzyme digests/breaks down gelatine/protein <u>and</u> releases the (dark) chemicals
(ii) *Any two from:*
Temperature of solution/trypsin
pH
Volume/depth of solution/trypsin
Size/length/area of film
Age/type/thickness of film/thickness of gelatin
Age of trypsin
(iii) Repeat the procedure <u>at each concentration</u> (and then calculate an average)
(iv)

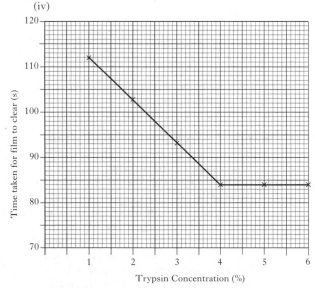

Trypsin Concentration (%)

Axes correctly drawn and labelled
Must have trypsin conc (%) **and** time for film to clear (s)
Points correctly plotted and line drawn
(v) There is <u>more</u> trypsin/enzyme (molecules)/active sites to react with the gelatine/substrate/protein
(vi) Surface area of film/size of film/thickness of gelatine is limiting the rate of reaction
or
The size of the film/gelatine is too small to allow all enzyme molecules to react with it
or
The reaction requires a minimum time to occur

(b) (i) The small intestine
(ii) So that they do not digest the cells/organs/pancreas/ glands/tissues <u>that produce them</u>
(iii) Vitamins/minerals/hydrochloric or stomach acid

5. (a) (i) X = SAN/SA Node/sino-atrial node/pacemaker
Y = AVN/AV node/atrio-ventricular node
(ii) The atria contract/atrial systole
(iii) Arrows must travel <u>down</u> the central wall of the heart from Y and <u>up each</u> side of the ventricles

(b) (i) Bicuspid/AV/atrio-ventricular
(ii) Ventricular systole

6. (a) Progesterone

(b) (i) (Causes the) repair/thickening/proliferation of the endometrium/lining
(ii) Stimulates/causes LH/FSH release / production

(c) Progesterone/hormone X remains high/constant/ does not decrease
or
Oestrogen remains high/does not decrease during the second half of the cycle/after day 24/25

(d) (i) P – Graafian follicle   Q – Corpus luteum
(ii) Ovulation/release of egg from ovary
or
surge in LH concentration

7. (a) Breathing rate remains constant <u>and</u> volume of each breath increases
Breathing rate remains constant at 14 breaths/min
or
Volume of each breath increases from 480 to 1240 cm^3

(b) 18

(c) 14 000

(d) (i) 1800 to 1840 <u>cm^3</u> (**units essential**)
(ii) Lung volume is nearing its maximum capacity
or
He is breathing as deeply as possible
or
Lungs have a limited capacity/can only hold so much air

(e) (Carbon dioxide is produced) by <u>respiration/the Krebs Cycle</u> (in body cells)

8. (a) <u>Three arrows drawn</u> – all pointing in the correct direction, ie:
hepatic artery into the liver
hepatic portal vein into the liver
hepatic vein out of the liver

(b) (i) Bile
(ii) Function –   Emulsification of lipids/fats
or
Emulsification correctly described – breakdown of large fat pieces into fat droplets
Explanation – This allows <u>enzyme/lipase to speed up</u> the breakdown (of lipids)
or
This <u>increases the surface area</u> (of lipids) for <u>enzyme/lipase</u>
or
Function –   Neutralisation of stomach acid
or
raises pH of intestine
Explanation – This provides the <u>optimum pH</u> for <u>lipase/enzymes</u>

(c) Glycogen/Iron/Vitamins (A or D)

9. (*a*) (The cerebrum) has a convoluted/folded surface/large surface
This allows for an increased number of cell bodies/cells/neurones

(*b*) Transfers/shares information/impulses <u>between</u> the two (cerebral) hemispheres/sides of the brain

(*c*) (i) The autonomic (nervous system)
(ii) Sympathetic speeds it up <u>and</u> parasympathetic slows it down

10. (*a*) 51 <u>weeks</u> (unit essential)

(*b*) 3, 4, 5 and 6

(*c*) *Any two from:*
Genes/inheritance
Encouragement/attachment
Diet
Environment
One has had an accident
One has had a disease/has a muscular disease
One has a slower myelination rate
One has a (physical) disability
One had a premature birth

(*d*) (i) Maturation
(ii) Myelination/development of myelin sheath (around nerve fibres)

11. (*a*) *Any two from:*
Use people of similar age/gender or gender balance/memory ability or span/use the same number of people/same first language

(*b*) <u>Short-term</u> memory/STM holds on average seven/5–9 words/items **or** capacity/span of STM
or
<u>Short-term</u> memory/STM can retain words for 30 seconds/a short time **or** duration of STM

(*c*) To prevent <u>rehearsal</u> of the words
or
To displace/remove the words from <u>short-term memory</u>

(*d*) 1. The meaning of words has no effect on their <u>recall/retrieval</u> from <u>short-term memory</u>
2. Related (meaning) words are harder to <u>recall/retrieve</u> from <u>long-term memory</u> (than unrelated words)
or
Unrelated (meaning) words are easier to <u>recall/retrieve</u> from <u>long-term memory</u> (than related words)

12. (*a*) (i) During Stage 2 it decreases <u>and</u> during Stage 3 it remains constant/steady/level
(ii) Rapid increase because <u>death rate</u> <u>drops quicker</u> than the <u>birth rate</u>
It levels off because <u>birth</u> and <u>death rate</u> become similar/equal
(iii) *Any two from:*
Increased/improved/better food supply/diet/agriculture
Increased/improved/better medical provision/vaccination/health care
Improved sanitation/hygiene/provision of clean drinking water

(*b*) (i) Pesticides remove (many) organisms/reduce species diversity/reduce biodiversity
or
Removal of pests/animals removes food sources for other species/organisms (further up the food chain)
or
Pesticides accumulate/build up along the food chain killing species/animals at the top of the food chain.

(ii) Selective breeding/genetic modification/genetic engineering/genetic manipulation/somatic fusion/crop rotation /irrigation/ mechanisation/ monoculture/ deforestation to <u>create agricultural land</u>/development of marginal land/ terracing / intensive farming

(*c*) (i) A large/exponential rapid increase in algae
(ii) *Any four from:*
1. Decomposition/decay (of dead algae by bacteria)
2. Increase in numbers of bacteria
3. Removal/decrease of oxygen (in the water)
4. Death of other species/fish/invertebrates/animals
5. Shading effect of algae leads to death of other plants
6. Toxic algae endangers other animals/man

SECTION C

1A (i) **Natural uptake and release of carbon**
1. (Carbon exists as) carbon dioxide in the atmosphere/air/water
2. <u>Photosynthesis</u> (by plants) takes up CO_2
3. Animals gain carbon by eating
4. CO_2 is released as a result of <u>respiration</u> (by living organisms)
5. <u>Decomposition/decay/breakdown by microbes/bacteria</u> releases methane/CO_2
6. Carbon becomes fossilised/forms fossil fuels/coal/oil/natural gas

(ii) **Disruption of the carbon cycle by human activities**
7. Burning/use of fuels releases carbon/ CO_2 (in the air)
8. <u>Increased population</u> has increased fossil fuel use
9. Industrialisation/transport uses (increased) fossil fuels/releases CO_2
10. <u>Deforestation</u> reduces photosynthesis/reduces CO_2 uptake
11. Increase in CO_2 in air causes <u>global warming/greenhouse effect</u>
12. Methane (CH_4) also causes global warming/is a greenhouse gas
13. Methane production caused by (increased) livestock farming/rice production
14. Domestic waste production/landfill creates methane

1B (i) **The role of neurotransmitters at the synapse**
1. The synapse/synaptic cleft is the junction/gap between neurones/nerve cells*
2. Neurotransmitters are stored in /released from vesicles*
3. Neurotransmitters are released on arrival of impulse
4. Neurotransmitters <u>diffuse</u> across the gap
5. Neurotransmitters bind with/reach <u>receptors</u>*
6. A threshold/minimum number of neurotransmitters is needed (for the impulse to continue)
7. Noradrenaline is removed by <u>reabsorption</u>
8. Acetylcholine is broken down by <u>enzymes/acetylcholinesterase</u>

(ii) **Converging and diverging neural pathways**
9. A converging pathway has several neurones linking to one neurone (if diagram must show direction of impulse)*
10. This increases the neurotransmitter concentration/chances of impulse generation
11. Any example of a converging pathway, eg <u>rods</u> of <u>retina</u>
12. A diverging pathway has one neurone linking to several neurones (if diagram must show direction of impulse)*

13. This means that impulses are sent to several destinations <u>at the same time</u>
14. Any example of a diverging pathway, eg fine motor control in <u>fingers</u> or release of sweat from <u>sweat glands</u>

* Can be given on **labelled** diagram

2A
1. Plasma is the liquid part of the blood
2. (*Any three from:*) named dissolved substances carried – oxygen, carbon dioxide, glucose, amino acids, urea, vitamins, minerals, etc
3. <u>Capillaries</u> have a large surface area/thin walls
4. <u>High pressure</u> (at the arterial end of the capillaries) forces fluid/plasma out
5. <u>Tissue fluid</u> (bathes the cells)
6. Plasma proteins/blood cells do not pass through capillary walls/stay in blood
7. (Dissolved) substances diffuse/move from tissue fluid into body cells
8. Waste products/named example diffuse/move out of the cells
9. <u>Low pressure</u> (at the venous end of the capillary network) allows return of fluid
10. Liquid/water also returns by <u>osmosis</u> (into the plasma)
11. (Excess) tissue fluid enters lymph vessels/lymph
12. This lymph/fluid is carried back to the blood (by lymphatic system)

2B
1. <u>Hypothalamus</u> detects/controls body temperature
2. (Thermo) <u>receptors</u> in the skin/body detect temperature
3. Temperature is maintained by <u>negative feedback</u> (mechanisms)
4. (Increased) sweating results in heat loss by <u>evaporation</u>
5. Increased blood flow to skin/vasodilation causes increased heat loss **or** reduced blood flow to skin/vasoconstriction reduces heat loss
6. <u>Arterioles</u> (not capillaries) constrict/dilate
7. Contraction of hair <u>erector muscles</u> makes hair stand up
8. This traps a layer of air which insulates/reduces heat loss
9. Increased metabolic rate causes heat production **or vice versa**
10. Adrenaline/thyroxine release occurs (when body is cold)
11. Shivering increases/causes heat production by <u>muscles</u>
12. Mechanisms are impaired in older people/undeveloped in infants

Hey! I've done it

Published by Bright Red Publishing Ltd, 6 Stafford Street, Edinburgh, EH3 7AU
Tel: 0131 220 5804, Fax: 0131 220 6710, enquiries: sales@brightredpublishing.co.uk,
www.brightredpublishing.co.uk

Official SQA answers to 978-1-84948-292-9
2008-2012